SIGNOR MARCONI'S
MAGIC BOX

GAVIN WEIGHTMAN, a writer and film-maker, is the author of *London River*, a history of the Thames, two volumes of *The Making of Modern London* (with Steve Humphries), *Bright Lights, Big City* and *The Frozen Water Trade*. He lives in London with illustrator Clare Beaton and Jack, Kate and Tom.

'Gavin Weightman's impeccably researched book is far more than a fact-led shunt through the Marconi story. His prose shimmers with the kind of romance that, in the mobile phone age, is quite difficult to grasp. But what a lovely story! An unassuming young chap confronts and defies the finest scientific minds in the world. It is pleasing to report that the cinematic aspect of this tale comes gloriously alive within Weightman's evocative, vividly detailed writing. Utterly captivating and, even for techno-dunces like myself, wholly illuminating'

Manchester Evening News

'A remarkable and compelling story'

Telegraph (Belfast)

D1409652

By the same author

The Making of Modern London: 1815–1914
The Making of Modern London: 1914–1939
London River
Picture Post Britain
Rescue: A History of the British Emergency Services
The Frozen Water Trade

Signor Marconi's Magic Box

How an Amateur Inventor
Defied Scientists
and Began the Radio Revolution

GAVIN WEIGHTMAN

HarperCollins*Publishers*

HarperCollins*Publishers*
77–85 Fulham Palace Road,
Hammersmith, London W6 8JB

The HarperCollins website address is:
www.harpercollins.co.uk

This paperback edition 2004

First published in Great Britain
by HarperCollins*Publishers* 2003

ISBN 978-0-00-713006-1

Set in PostScript Linotype Adobe Caslon
with Bauer Bodoni and Spectrum Display

For my father

CONTENTS

LIST OF ILLUSTRATIONS   ix

ACKNOWLEDGEMENTS   xiii

INTRODUCTION   xv

MAP: MARCONI'S EARLY WIRELESS
     TELEGRAPH STATIONS   xviii

1   In Darkest London   1

2   Silkworms and Whiskey   10

3   Sparks in the Attic   16

4   In the Heart of the Empire   21

5   Dancing on the Ether   25

6   Beside the Seaside   35

7   Texting Queen Victoria   39

8   An American Investigates   44

9   The Romance of Morse Code   49

10   A New York Welcome   58

11   Atlantic Romance   66

12   Adventure at Mullion Cove   71

13   An American Forecast   82

14   Kite-Flying in Newfoundland   88

15   The Spirits of the Ether   93

16   Fishing in the Ether   100

17   The End of the Affair   108

18   Farewell the Pigeon Post   115

19 The Power of Darkness   121
20 The Hermit of Paignton   128
21 The King's Appendix   132
22 The Thundering Professor   142
23 A Real Colonel Sellers   151
24 Defeat in the Yellow Sea   161
25 A Wireless Rat   167
26 Dazzling the Millions   172
27 'Marky' and his Motor   178
28 On the American Frontier   186
29 Marconi gets Married   191
30 Wireless at War   198
31 America's Whispering Gallery   202
32 A Voice on the Air   206
33 The Bells of Budapest   211
34 Wireless to the Rescue   219
35 Dynamite for Marconi   225
36 *Le Match Dew–Crippen*   230
37 A Marriage on the Rocks   236
38 Ice and the Ether   242
39 'It's a CQD, Old Man'   247
40 After the *Titanic*   253
41 The Crash   260
42 The Suspect Italian   268
43 Eclipse of Marconi on the Eiffel Tower   277
44 In Bed with Mussolini   281

EPILOGUE   289
INDEX   293

ILLUSTRATIONS

All photographs © Marconi plc unless otherwise indicated.

Guglielmo Marconi shortly after his arrival in London in 1896. *(Science Museum/Science & Society Picture Library)*

Marconi, aged about four, with his mother Annie and his brother Alfonso.

The Villa Griffone.

The little 'coherer' which began the radio revolution. *(Science Museum/Science & Society Picture Library)*

William Preece.

One of the 'magic boxes' Marconi used to make his first demonstrations of wireless.

Marconi demonstrates his wireless system to observers from the Italian navy at La Spezia in 1897.

British Post Office engineers testing Marconi's spark transmitter in 1897.

The world's first working wireless station, at the Royal Needles Hotel on the Isle of Wight.

The *Daily Mail*'s illustration on 19 August 1898 of a Marconi engineer tapping out Morse code messages from the Isle of Wight to the royal yacht *Osborne*. *(British Library Newspaper Library)*

One of the first wireless 'text messages' in history, dictated by Queen Victoria on the Isle of Wight.

George Kemp and Marconi in their temporary wireless station at Wimereux on the northern coast of France in 1898.

The Haven Hotel, Poole, Dorset, Marconi's home and research station for several years.

Reginald Fessenden at his research station at Brant Rock, Massachusetts. *(George H. Clark Radioana Collection, Archives Center, National Museum of American History, Smithsonian Institution)*

Ambrose Fleming.

The *Transatlantic Times*, the news-sheet produced by Marconi on the SS *St Paul*.

A love letter from Josephine Holman to Marconi, partially in Morse code.

Marconi watches as helpers attempt to get a huge kite airborne at St John's, Newfoundland, in December 1901.

A chart confirming the distances Morse messages were received as the SS *Philadelphia* liner sailed from Southampton to New York in February 1902.

A 'magnetic detector' devised by Marconi in 1902.

Marconi and Sir Henneker Heaton MP celebrate Edward VII's coronation in 1902.

Lee de Forest. *(Bettmann/CORBIS)*

An exhibit put on at the St Louis World's fair in 1904 by Abraham White. *(George H. Clark Radioana Collection, Archives Center, National Museum of American History, Smithsonian Institution)*

The gigantic power packs of Marconi's transmitter at Clifden, on the west coast of Ireland.

The final scene of Inspector Drew of Scotland Yard's pursuit of Hawley Harvey Crippen across the Atlantic.

A message sent from the liner *Virginian* to the *Californian* at 4 a.m. on 15 April 1912, two hours after the *Titanic* sank.

American cartoon celebating Marconi's role as the hero of the *Titanic* disaster.

Marconi with his wife Bea and their three children, Degna, Giulio and Gioia, around 1920.

Oliver Lodge. *(Barratts Photo Press Ltd)*

The mobile Marconi wireless kit used by British soldiers in the First World War.

Marconi with his second wife Cristina Bezzi-Scali on the yacht *Elettra. (Science Museum/Science & Society Picture Library)*

ACKNOWLEDGEMENTS

My greatest debt is to Louise Jamison, archivist of the Marconi
Company, who found the time to unearth for me many newspaper
cuttings, diary extracts and illustrations and who made me welcome
at Chelmsford, where the records going back to 1896 are kept.
Gordon Bussey, the company's historical consultant, provided
much technical assistance, in particular on the details of the first
transatlantic signal of 1901. At the Villa Griffone, Barbara Valotti
made me welcome, and Maurizio Bigazzi demonstrated some of
Marconi's very early equipment and showed me the lie of the land.
Professor Giuliano Pancaldi of Bologna University gave me some
valuable background information and a handsome book, some of
which was translated for me by Mariluisa Achille. The only really
intimate account of Marconi's life is in his daughter Degna's
memoir *My Father, Marconi*.

A very special thanks to Thomas H. White, an American I have
not met but whose personally compiled website 'United States
Early Radio History' contains an astonishing range of original
material in the form of newspaper and magazine articles dating
back to the late nineteenth century, many of which are not available
in Britain. As always the British Library and the London Library
were invaluable, and I would like to thank Kate Simpson and
Patrick McDonnell for their diligent digging in the newspaper
archives at Colindale.

Susan J. Douglas's wonderful academic work *Inventing American
Broadcasting* was an inspiration early in my research, and Mary K.
McCleod's *Marconi: The Canada Years* provided a valuable account
of the first wireless stations in Nova Scotia. Brian Stewart in
Toronto and Clare Beaton in London made some very pertinent

comments on early drafts of the book. At HarperCollins, Richard Johnson's enthusiasm for the project was invaluable and Robert Lacey's editing as meticulous and astute as always. I would like to thank them and my agent Charles Walker at Peters, Fraser and Dunlop for all their assistance with the book.

The history of early wireless is fraught with claims and counter-claims about who really invented what, and I have done my best to take an objective view. Any mistakes I have made are naturally my responsibility alone.

Gavin Weightman
London, September 2002

INTRODUCTION

It was the most fabulous invention of the nineteenth century. The public and the popular newspapers regarded it as nothing short of miraculous, and the leading scientists of the day in Europe and America, whose discoveries had made it possible, could not understand how it worked. Wireless in its pioneer days had nothing to do with home entertainment: no speech or music was transmitted. But the tap, tap, tap of the telegraph key spelling out messages which had travelled mysteriously through the 'ether' was exciting enough in a world still mostly horse-drawn and coal-fired, a world without cinema or the motor-car, in which the telephone was still an expensive luxury and great cities like London and New York had only recently winced at the brightness of electric light.

The wider world first learned about the possibilities of wireless telegraphy, as the new invention was called, in 1897. In November that year the very first wireless station was established in the exclusive Royal Needles Hotel. This splendid clifftop Victorian pile took its name from the nearby pillars of eroded rock which jut into the sea on the western corner of the Isle of Wight, a short ferry-ride off the south coast of England. An odd assortment of wires strung out on posts tethered to the ground to secure them against the stiff sea breezes and the not infrequent gales was the only outward sign that mysterious signals were being sent to the steamers, packed with skylarking holidaymakers, which plied the coast. Guests at the Royal Needles were aware when the transmitter was in operation, for they could hear and sometimes see the crack of electrical sparks activated by a Morse code key pressed by one of the operators inside the hotel. The range of these signals was only a few miles. But the fact that it was possible at all to establish, by remote control, communication with a ship steaming along at a rate of knots, *even when it was lost to view behind a cliff*, was nothing short of astonishing. The wonders of science, it seemed, would never cease.

Only the year before, the newspapers had been full of accounts of the 'new photography' called X-rays, which could 'see' through solid objects. The public now had to take in the amazing possibilities presented by the 'new telegraphy'. Like most powerful new inventions, wireless had the potential to bring both good and evil to the world: could it be used, perhaps, as a weapon? Might an electric wave aimed at a battleship blow up its explosive magazine as surely as any shell from a shore battery?

This 'new telegraphy' was not only mind-bending in its apparent defiance of contemporary scientific understanding; there appeared to be a very real prospect of it completely replacing the network of telegraphic cables which in the previous half-century had been strung out over land and laid across the ocean beds, at colossal expense. At the very least it meant that ships at sea, including the great liners which were then carrying millions of European immigrants to North America, need never be out of contact with each other and with New York, Liverpool and London. The big question was: how far could these invisible waves travel through the 'ether', carrying Morse-coded messages which were decipherable at a receiving station?

In 1897, nobody had the answer to that question. The great majority of physicists who worked on what were known as 'Hertzian waves' very much doubted that they could be used for communication over distances greater than a mile or two. Even that range, which had already been achieved, was causing some puzzlement, for it was not known through what medium the waves from a wireless transmitter really travelled. Did they go through hills, or over them? Did they bend around the curvature of the earth's surface? As they were akin to light and travelled at the same speed, why did they not simply dissipate into the atmosphere, never to be retrieved and decoded on the ground?

Though there were some ingenious speculations about how wireless waves travelled long distances there was no definitive answer until the 1920s, by which time radio had become a sophisticated industry, filling the airwaves with a cacophony of sound – much of it American. In the meantime, from 1897 until the cataclysm of the First World War, wireless telegraphy was woven into the social and economic fabric of the most sophisticated societies with astonishing speed.

This is the story of how one of the most extraordinary inventions in history came about. Taking the leading role in a cast of many brilliant and eccentric characters is Guglielmo Marconi himself, whose home-made magic box first brought the 'new telegraphy' to the notice of the general public. In his lifetime he enjoyed worldwide fame for the achievement of turning a boyhood fascination with electricity into an entirely new form of communication, and a huge industry. Marconi was one of the greatest amateur inventors of all time. It is remarkable testimony to the fragility of reputation that a man who could command such respect in his lifetime should now be relegated to comparative obscurity, and that the names of scores of his contemporaries who made radio work have no resonance at all for a generation addicted to the most modern form of wireless telegraphy: text messaging on a mobile phone. That Queen Victoria received text messages sent by wireless from the royal yacht to her home on the Isle of Wight more than a century ago will come as a surprise to those who imagine the technology of the mobile phone is almost brand new. The story begins way back in the days of dark streets, horse-drawn carriages, and the blood-curdling murders of Jack the Ripper.

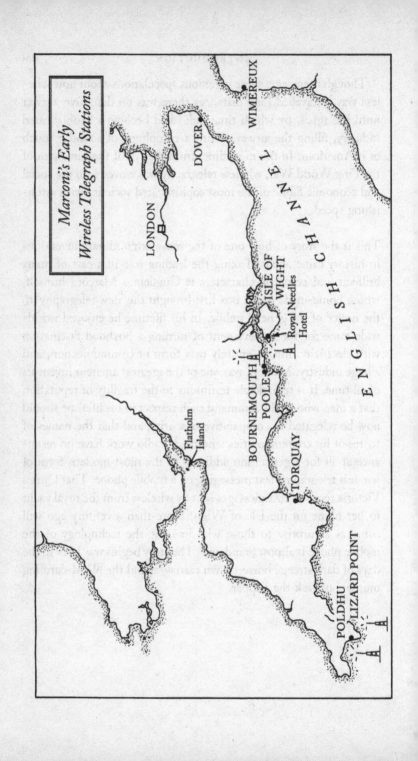

Marconi's Early
Wireless Telegraph Stations

ENGLISH CHANNEL

WIMEREUX

DOVER

LONDON

ISLE OF
WIGHT

Royal Needles
Hotel

BOURNEMOUTH

POOLE

Flatholm
Island

TORQUAY

POLDHU
LIZARD POINT

1

In Darkest London

On a winter's evening in 1896 a brougham, a four-wheeled cab drawn by a single horse, left the fashionable stuccoed terraces of west London and headed eastwards along the dirt roads and cobbled streets of the capital, which glistened in the gaslight under a light rain. The passengers were a young man who had with him two large black boxes, and a gentleman in his sixties sporting a long grey beard, his thinning hair pasted to his head in a centre parting. Steam rose from the horse's flanks in the dank air as the brougham rattled through the canyons of streets in the City and, leaving the Square Mile, came to the fitfully-lit roads of Whitechapel. This was the frontier of the notorious East End, where only a few years earlier Jack the Ripper had mutilated his victims and left them dead or dying in dark alleyways.

The cab turned onto Commercial Street, and the young man peered through the smoke-filled air for a sight of their destination. Finally they lurched off the main road and entered a courtyard fronting an elegant building that looked as if it might have stood there for hundreds of years. This was Toynbee Hall, which had in fact been completed just fifteen years previously. It was the inspiration of the remarkable Canon Barnett, vicar of the poverty-stricken parish of St Jude's in Whitechapel, who had chosen to conduct his missionary work not in Africa but in that part of the capital William Booth, founder of the Salvation Army, had called

'Darkest London'. Toynbee Hall, modelled on Oxford and Cambridge's colleges, was a 'settlement' built with money subscribed by those ancient universities. Here some of the leading figures of the coming generation of politicians and civil servants were invited to live for months at a time, so they could learn about poverty and offer some culture and instruction to the poor. There was a large lecture theatre, in which many distinguished people had delivered their opinions on the great moral, political and scientific issues of the day. A few years later the Russian revolutionary Vladimir Ilyich Lenin would attend lectures at Toynbee Hall.

The speaker this evening, Saturday, 12 December 1896, was not the young man who unloaded his black boxes from the cab: he and his apparatus were to be the star turn of the lecture which was to be given by his older companion. The two had met for the first time only that April, and the bearded Victorian gentleman had subsequently been so impressed by the young man's invention that he had become his patron. Some private demonstrations of what the black boxes could achieve had been given on the rooftops of London and out on the open chalk lands of Salisbury Plain, where the British Army rehearsed cavalry charges close to the Neolithic monument of Stonehenge. But this evening at Toynbee Hall was to be the first exhibition of the magic boxes to a public audience. The lecture was entitled 'Telegraphy without Wires', a subject about which little was then known outside the science laboratory and the telegraph business itself.

Toynbee Hall was packed. The speaker, William Preece, had gained a reputation for delivering lucid and amusing public lectures on recent exciting scientific discoveries. On this evening he did not at first reveal who his accomplice was, but gave a little of the history as he knew it of methods of sending telegraph messages without a wire connection. As long ago as 1838 a German, Professor Steinbjel, one of half a dozen scientists who claimed to have invented the electric telegraph, had foreseen a time when it might be possible to do away with the cable altogether.

In fact, Preece continued, he himself had already achieved this.

Just two years previously he had been astonished to discover that messages being sent on underground telegraph cables owned by the British Post Office could be picked up by the exchange of a telephone company in the City, which had its wires above ground. Somehow the electronic impulses in one wire had jumped across to another, creating, in effect, a form of 'wireless' communication. Some experiments had been carried out to see if this could be the basis of a new system of communication. Some limited success had been achieved, but that evening Preece had an important announcement to make about an entirely new form of wireless telegraphy. It was at this point, according to newspaper reports which appeared the following Monday, 14 December, that Preece introduced his audience to the young man who shared the platform with him. He was an Italian electrician named Guglielmo Marconi who, Preece explained, had come to him recently with his home-made equipment. This evening he and Signor Marconi would for the very first time demonstrate to a general audience the working of this system.

'The apparatus was then exhibited,' said the *Daily Chronicle* report. 'What appeared to be just two ordinary boxes were stationed at each end of the room, the current was set in motion at one end and a bell was immediately rung in the other. To show there was no deception Mr Marconi held the receiver and carried it about, the bell ringing whenever the vibrations at the other box were set up.' When Preece pressed a lever in the sending box there was the crack of an electric spark, and an instantaneous ringing in the receiver held by Marconi. The effect was achieved, the audience was told, by the transmission from the sending box of 'electrostatic' waves much the same as light. These were received by the other box, in which there was a device which, when activated, rang the bell. In other words, a signal was being sent around the lecture hall which was invisible, but as tangible in its effects as any telegraph impulse sent along a wire. And it followed Marconi wherever he went in the hall.

To any modern audience this device would look more like a

mildly diverting toy than an invention at the very forefront of technology. No transmission of speech, or music, or anything now associated with radio was being demonstrated. No messages were being sent at all – just an invisible electronic signal. But in 1896 that was sensational enough. It was like some fantastic act at the music hall. In fact, those present might easily have dismissed the demonstration as the work of a magician and his assistant, for the young man had a suspiciously exotic Italian name, although he looked and talked like a smart Londoner about town. However, there could be no doubts about the credentials of the speaker: the sixty-two-year-old William Preece, shortly to become Sir William, was Chief Electrical Engineer of the single most powerful communications system in the world, the government-owned British Post Office.

Only a handful of people in London had heard of the twenty-two-year-old Guglielmo Marconi. He said a few words to the audience in his impeccable, slowly enunciated English when the demonstration was over. Without the authority given it by William Preece's presence, the lecture would probably have had little impact, and the audience would have climbed back into their cabs and carriages muttering about the devious sleight of hand of foreigners. But Preece assured them that he had seen a number of demonstrations of this young man's method of transmitting signals, and that it held out the very real prospect that, with some modifications, it would be able to send messages through the ether over distances of several miles. Marconi's wireless waves could activate a Morse code printer, producing an instant and invisible means of conveying exactly the same kind of messages that were then being tapped in dots and dashes around the world on the global cable telegraph network.

How many in the audience that night realised that they were seeing history in the making, we do not know. Preece, however, appeared to be full of confidence about its potential. He pledged the Post Office's support for the development of Marconi's invention, and dismissed as irrelevant the claims made for an Indian,

Professor Jagdish Chandra Bose,* as the true discoverer of wireless telegraphy. There was loud cheering when Preece told the audience that what had been demonstrated that evening would give Britain's mariners 'a new sense and a new friend', and would make navigation infinitely easier and safer than it then was.

Preece flicked open the cover of his gold hunting-watch, and drew the lecture to a close. That evening he took young Marconi back across the city to a house rented in fashionable Westbourne Park before going on to his own home in Wimbledon, eight miles out of London.

Though they were very different in age and background, there was clearly an affinity between the two men. Both, in their different ways, had disliked formal education, but were fanatical workers when a subject interested them. Preece was dismissive of academics, who were always claiming superior knowledge of physics and the mysterious workings of electricity but produced nothing of practical value. Writing about his own childhood, he had said that boys always rebel against their fathers and learn only from their mothers. At least, that had been his experience, for he owed all his success to his Welsh mother. That night, after his first successful public demonstration of his magic boxes, it was Marconi's mother who was there to greet the young man on his return home. Preece had become his patron, but his Irish mother Annie Jameson had championed him since childhood, giving him emotional support and encouragement, while his sterner Italian father had paid the bills for his son's experiments. And it was his mother who, through the connections of her wealthy and influential family, had managed to arrange the fortuitous meeting between her son and the distinguished head of the British Post Office's engineering operations.

William Preece was old enough to recall the invention of the safety match for lighting home fires in his native Wales, and had

* In 1895 Professor Bose, of the Presidency College, Calcutta, had succeeded in ringing a bell and exploding a mine with electro-magnetic waves while working along the same lines as Marconi.

spent his working life experimenting with, adopting and adapting the new electronic technologies as they were revealed to the world. Twenty years earlier he had toured the United States and had met Thomas Edison, America's most celebrated inventor, who served him raw ham, tea and – to his astonishment – iced water in summer.

As well as enjoying Edison's chilled drinks, Preece had been one of the first to try out Alexander Graham Bell's telephone, and had brought the equipment back to England, where it was an object of incredulous fascination. Could you actually recognise the voice of someone on the other end of the line, people wondered. When the telephone was still at the development stage, conversations sounded like a high-pitched exchange between the protagonists of a seaside Punch and Judy show. Preece had initially regarded Bell's invention as no more than a 'scientific toy'. Now he had his own phone number at the Post Office headquarters in St Martin's-le-Grand, and the telephone was no longer a novelty. But there were new inventions to startle the public. In the very week that young Marconi had arrived in London early in 1896 he had read in the newspapers of an astonishing discovery. A Bavarian physicist, Wilhelm Conrad Roentgen, had chanced upon a way of 'photographing the invisible' with mysterious rays produced by electricity passing through a vacuum. The ability to see through solid objects was the stuff of science fiction, yet Roentgen had produced a photographic image of the bone structure of a human hand. As he did not know what the electric waves were, he called them 'X-rays'.

News of Roentgen's amazing discovery had broken on 5 January 1896 in a Vienna newspaper, and had rapidly been telegraphed around the world. There was much chatter about the danger of X-rays to the modesty of women: wicked inventors might be able to see through their clothing. Scientific discovery was often frightening as well as exciting, and the penetrating powers of X-rays were the subject of much anxious debate, though nobody then knew about the dangers of radiation. The issue was privacy.

The publicity William Preece had afforded Marconi very quickly drew the attention of newspapers and magazines, and when news of his 'Marconi waves' began to spread the public were intrigued to know if they too might threaten the privacy and decency of English ladies. After all, his magic boxes sent and received invisible signals which could apparently travel much further than Roentgen's X-rays. Young as he was, Marconi found himself called upon to provide extensive interviews. Just three months after the Toynbee Hall lecture, in March 1897, the *Strand Magazine* published an article by H.J.W. Dam with the title 'The New Telegraphy'. It was syndicated worldwide by the enterprising American magazine *McClure's*.

Dam had been to see Marconi at his home in Westbourne Park in the hope of learning something about this young man whose discoveries were 'more wonderful, more important and more revolutionary' than Roentgen's 'new photography'. He found himself greeted by a most unusual character, who was 'completely modest' and made no claims at all as a scientist. This 'tall, slender young man', who looked at least thirty, had a 'calm, serious manner and a grave precision of speech' which gave the impression that he was much older than he was. Speaking in his 'perfect' English, he told the reporter that he had been for ten years an 'ardent amateur student of electricity'.

In the calm, considered manner which was to be his hallmark whenever called upon to explain his discoveries to the public, Marconi told Dam how he had found to his surprise while experimenting with electric waves on his father's country estate outside Bologna that he could generate signals which went through or over hills. He really had no idea how they got there, but he had proved over and again that a rise in the land three quarters of a mile across was no obstacle to the transmission and reception of these electronic signals. Marconi explained that he had begun by copying the laboratory equipment of the great German physicist Heinrich Hertz, and had adapted it so that he could send Morse messages. But whereas Hertz had sent his electro-magnetic waves only a few yards, Marconi had achieved much greater distances, and he was not sure if he had, by chance, discovered a previously unknown phenomenon: a new kind of 'wave'.

The science Marconi was working with was not well understood. In 1865 the Scottish physicist James Clerk Maxwell had proposed that electro-magnetic forces travelled in waves. These were analogous to sound and light waves, but could not be detected by the human ear or eye. They travelled at the speed of light, but were invisible, because the eye could only detect certain wavelengths. Maxwell's model was purely mathematical, and he left it to others to find a way of generating and measuring these waves. Hertz had been the first to achieve this, publishing his findings in 1888. He used a spark to generate the waves which he bounced back and forth in his laboratory. Crudely speaking, the size of the spark 'gap' determined the length of the waves, and Hertz had worked with fairly short waves. Marconi had experimented with a whole range of different spark transmitters, and had produced results which appeared to be substantially different from those of Hertz. In fact, because he believed his apparatus could produce waves that could reach parts impenetrable by those generated by Hertz, Marconi thought he might have chanced upon some new kind of electro-magnetic signal. Dam asked: 'What is the difference between these and the Hertz waves?'

Marconi replied: 'I don't know. I am not a professional scientist, but I doubt if any scientist can tell you.' He thought it might have something to do with the form of the wave. As to the nuts and bolts of his equipment, Marconi said apologetically that he could not say more because it was being patented, and was therefore top secret. What he could tell the astonished reporter was that his waves 'penetrate everything and are not reflected or refracted' even by solid stone walls or metal. He could even send them through an 'ironclad', a heavily reinforced battleship.

This last claim set up instant alarm in the reporter: its implications were far more serious than the possibility of X-rays compromising the modesty of ladies. 'Could you not from this room explode a box of gunpowder placed across the street in that house yonder?' Dam asked.

'Yes,' Marconi replied confidently. 'If I could put two wires or two plates in the powder, I could set up an induced current which would cause a spark and explode it.'

'At what distance have you exploded gunpowder by means of electric waves?'

'A mile and a half.'

Could Marconi's instruments ignite the explosive magazine of an ironclad and blow it up from a distance? Already the Royal Navy was concerned that if its ships carried wireless telegraphy equipment, the signals might blow up their own stores of powder. It could be a problem, Marconi conceded. Beams from electric lighthouses along the coast could destroy an unwary fleet in seconds. Warming to this notion, Dam wrote: 'Of all the coast defences ever dreamed of, the idea of exploding ironclads by electric waves from the shore and over distances equal to modern cannon ranges is certainly the most terrible possibility yet conceived.'

Blowing up ships, however, had never been in Marconi's mind. Quite the reverse. From boyhood, when his father had bought him a yacht which he sailed in the Bay of Genoa, he had loved the sea. Though he had no clear idea how his wireless waves would be used in practical terms, he did imagine that there was a real prospect of communications between ships and shore, and between ships on the open ocean, where there were no telegraph cables.

In London, Marconi and his mother could have enjoyed a glamorous social round: nightly balls, dinner parties, the opera, Ascot and all the trappings of the Season. Annie had many relatives in town, and always enjoyed her trips to England. But there was to be little time for frivolous socialising: Guglielmo had succeeded beyond their wildest dreams, and he was fearful that if he did not move fast someone else would overtake him in the exploitation of the new telegraphy. After all, he was only an amateur whose invention was homespun, devised after long hours working alone in the attic of the family country home in Italy.

2

❧❧❧❦❦❦

Silkworms and Whiskey

The Villa Griffone is set in its own grounds of orchards, vine-yards and fields in rolling countryside outside the village of Pontecchio, near Bologna. Bologna had water-powered silk-weaving mills long before the Industrial Revolution transformed British industry in the late eighteenth century. The city had a distinguished history of scientific discovery, and was the home of the eighteenth-century pioneer of electrical forces Luigi Galvani. Nearly all the advances in the study of electro-magnetism had been achieved by trial and error, in the absence of any useful theory. In fact theory had sometimes got in the way of understanding, as is often the case. Galvani, a Professor of Anatomy at the ancient University of Bologna, had come to the conclusion that frogs could produce electricity after the chance discovery that specimens he was dissecting reacted to an electrical current. His disciple, and later his opponent, Alessandro Volta showed that the frogs were merely acting as crude batteries, and went on to create the first means of storing electricity which could be tapped for a continuous source of current. The work of both men has been commemorated in the terms 'galvanised' and 'voltage'.

From an early age Guglielmo Marconi was familiar with Bologna's scientific heritage, and in the long summer days at the Villa Griffone he began his first experiments with the mysterious forces of electricity. Marconi's heritage – and the pioneer days of wireless – arose from a most unlikely union between Irish whiskey and Italian silkworms. That his mother and father should have met at all was remarkable, that they should have fallen in love even more so, and that they married in the teeth of opposition from her family the most unlikely event of all. Theirs was a story of high romance, yet precious little of it is known apart from the reminiscences of Marconi's mother, recorded much later by her granddaughter Degna.

Annie Jameson was born in 1843, one of four daughters of Andrew Jameson of County Wexford in Ireland, the well-known and wealthy distiller of Jameson's Irish whiskey. The family lived in an old manor called Daphne Castle, which had parkland and a moat. Annie had one outstanding talent: singing. As a teenager she had wanted to perform in opera, and according to the family legend had been invited to sing at the Royal Opera House in Covent Garden. Her father refused to let her go: the stage was not in those days regarded as a suitable place for well-bred young ladies. As compensation for the thwarting of her ambitions it was arranged that Annie should go to Bologna to study singing. There she could stay with business contacts of the Jamesons, a respectable Italian family called de Renolis, and could sing to her heart's content without risk to her family's reputation.

The de Renolis family had suffered a personal tragedy a few years before Annie arrived to stay with them. In 1855 their daughter Giulia had married a moderately prosperous landowner called Giuseppe Marconi. In the same year Giulia had given birth to her first child, a son, Luigi. Sadly, as happened so often at that time, the young mother survived the birth of her son by only a few months. Giuseppe, now a lone parent, remained close to the de Renolis family. He had moved to Bologna from the hill country of the Apennines, which run like a backbone through north central Italy. When his wife died he asked his father, who still lived in the mountain village where Giuseppe had been brought up, to join him in Bologna. The ageing Domenico Marconi agreed, sold up his mountain estate and moved to the city. But Bologna was too busy and confined for him, so he bought an estate at Pontecchio, eleven miles away. In the large, square, plain but handsome Villa Griffone he took to raising silkworms, and made some success of it, while his widower son Giuseppe husbanded the orchards and the fields in the rolling countryside.

When Annie Jameson came to stay with the Renolis she was introduced to their bereaved son-in-law and little grandson Luigi. Giuseppe lived more at the Villa Griffone than in Bologna, and Annie must have spent some time there too, for she fell in love with the place and with him. She returned to Ireland to ask her family for permission to marry her Italian sweetheart, but they flatly refused to consider it. According to her granddaughter Degna, the grounds for rejection were that he was much older than her (by about seventeen years), he already had a son, and to top it all he was a foreigner. Annie had to bow to the authority of her father, and appeared to accept the decision. But she kept in touch with Giuseppe, with letters somehow smuggled between Ireland and Italy, and vowed to run away to marry him when she reached the age of majority at twenty-one. This she did, meeting him in Boulogne-sur-Mer on the northern coast of France, where they married on 16 April 1864. As husband and wife they took stage coaches across France, over the Alps and back to Bologna and the Villa Griffone. Their first child, Alfonso, was born a year later. Nine years later, in April 1874, Annie gave birth in Bologna to a second son, Guglielmo. Both boys were baptised Roman Catholic, although their mother was Protestant.

Giuseppe had no family other than his in-laws. His father had died, and a brother who became a priest had been murdered by a thief. Annie, on the other hand, had three sisters, all of whom had married and had children. Annie did not lose touch with them despite her elopement. One of her sisters had married an English military man, General Prescott, who was posted to Livorno on the north-west coast of Italy. Annie often took Alfonso and Guglielmo to stay with her, where they enjoyed the company of the Prescotts' four daughters and the small English community.

The English girls were also often Guglielmo's playmates at the Villa Griffone, and he spent so much time with them and with his mother that at times Italian became his second language.

Annie read the Bible to her sons as part of their English lessons, and appears to have had no interest in science. Of greater interest to Guglielmo than religious instruction was the library in the Villa Griffone, which contained a wide selection of books. It is not clear whether it was Guglielmo's father or his grandfather who had collected works ranging from Thucydides' *History of the Peloponnesian War* to the lectures of the brilliant English chemist Michael Faraday. Perhaps many of the scientific works were provided by or for the young Guglielmo himself; in any case, from the age of about ten Guglielmo began to work his way through this store of knowledge. He became especially interested in the wonderfully lucid lectures of Faraday, who had made some of the most significant discoveries about the relationship between electricity and magnetism, and had invented the first, tiny, electrical generator.

Born in 1791, the son of a blacksmith, Faraday had had no formal education, and began his working life as an apprentice to a bookbinder. He had attended a series of lectures given by the famous scientist Humphry Davy at London's Royal Institution, made notes on them and sent them to Davy, asking for a job as a laboratory assistant. Davy took him on, and eventually Faraday was to succeed him as the most celebrated scientist in England, spending his life experimenting in a variety of fields, but most significantly on the nature and applications of electricity. He died in 1867, just seven years before Guglielmo Marconi was born. Faraday undoubtedly provided the young Marconi with a heroic model: the scientist alone in his laboratory with wires and chemicals, painstakingly testing his theories. But the greatest hero to descend from the shelves of the Villa Griffone library was the American Benjamin Franklin. Among the many achievements of this extraordinary man, born in 1706, a printer, diplomat and amateur scientist who was at seventy the oldest signatory of the Declaration of Independence, was the invention of the lightning conductor.

In a celebrated experiment, Franklin had flown a kite in a thunderstorm to demonstrate that the electrical charge of lightning could be channelled along a wire to which the kite was tethered. This clearly impressed the young Guglielmo, for his daughter recalls him telling the story of how he and a friend rigged up a lightning conductor in the house they were staying in in Livorno, and prayed for a storm. When one came they were thrilled to discover that their toy worked: at every lightning flash, the electrical charge triggered a little mechanism which rang a bell in the house. A replica of young Guglielmo's lightning alarm is among a wonderful collection of his early gadgets in the Villa Griffone, which is now a museum devoted to his extraordinary childhood inventiveness.

It was around the time of the lightning experiment, in 1887, when Marconi was thirteen years old, that the German scientist Heinrich Hertz made known his discovery of electro-magnetic waves, prompting the Irish mathematician George Fitzgerald to declare that humanity had 'won the battle lost by the giants of old . . . and snatched the thunderbolt from Jove for himself'. This was a humbling statement for Fitzgerald to make, for only a few years earlier he had announced that he believed the artificial creation of electro-magnetic waves was not possible, thereby blunting the ambition of British scientists working along the same lines as Hertz.

Guglielmo did have some academic tutoring at an institute in Livorno and a college in Florence, but his serious work was carried out on his own at the Villa Griffone. He was privileged, for his father not only provided him with a library, but grudgingly subscribed to all the leading scientific journals of the day, which Guglielmo devoured. His boyhood notebooks, rediscovered in Rome only seven years ago, are testimony to his fanatical interest in electricity and all the latest theories and inventions. The scientific community was most excited at the time by the work of Hertz. His apparatus for proving the existence of the Scottish physicist James Clerk Maxwell's imagined electro-magnetic waves and measuring their 'length' was quite crude. A spark was produced by jumping electricity across a gap between two metal balls charged by Leyden jar batteries. The spark generated electronic waves which travelled invisibly across Hertz's laboratory to activate a 'receiver' made up of wires which produced a spark in response. His experiments inspired many other scientists to examine the properties of what became known as Hertzian waves.

In 1894 Heinrich Hertz died at the tragically young age of thirty-six. During an operation for cancer of the jaw he suffered blood poisoning, which killed him. The scientific magazines were filled with obituaries which gave accounts of the trail-blazing experiments he had conducted. When the young Marconi read these he at once conceived the idea of using the apparatus which Hertz had made to send telegraph messages. He did not know it at the time, but precisely the same idea had struck a number of scientists and inventors in England, America and Russia.

A neighbour of the Marconi family at Villa Griffone was the Italian Professor of Physics Augusto Righi, who had done his own work on Hertzian waves. Guglielmo was thus able to discuss his idea with a leading scientist. He received little or no encouragement, but quite probably he managed to get an idea of how to construct the kind of transmitter and receiver Hertz had used in his laboratory, and with the help of his mother he cleared out an area on the upper floor of the Villa Griffone which had been used by his grandfather for keeping silkworms. This home-made laboratory has been lovingly restored by the staff of the modern museum. The beautifully recreated models of his early equipment are testimony to Marconi's skill, and his dedication to an ambition on which he spent nearly all his waking hours.

By the time Marconi was a teenager there was a widespread interest in electricity, which was reflected in the publication of a range of journals from which the enthusiastic amateur could learn about the very latest theories and discoveries made in Europe and America. The majority of these were in English, and Marconi's easy command of the language ensured that there were few developments of which he was unaware.

3

❧❧❧❧

Sparks in the Attic

Day after day through the hot summer months of 1895, Guglielmo Marconi climbed the stairs to his makeshift workshop in the attic of the Villa Griffone. He said very little to his family about what he was trying to achieve behind its closed door. Early on he learned to be cautious about making any predictions, and he was very conscious of his ageing father's view that the whole thing was a waste of time. Being a scientist or an inventor was not, in Giuseppe Marconi's opinion, a 'career', unless, like their neighbour Righi, you had a professorship.

From time to time Guglielmo would allow his English cousins to visit the attic, where he would show them the magic he could perform with crackling sparks which made a bell ring by a mysterious force. He himself could not really explain how these tricks worked. He achieved them by trial and error, making use of every bit of electrical equipment and every published experiment he could lay his hands on. For his electricity supply he could buy batteries. It was also a simple matter to get hold of a Morse key and a Morse printer, for these were mass-produced for the telegraph industry, and there were many models on the market.

Morse code was a set of dots and dashes which represented the letters of the alphabet. The sender pressed a lever on the key, making an electrical connection which in turn activated a circuit connected to a printer which recorded either a dot or a dash. Hold the lever down for a short time, and it was a dot; longer, and it was a dash. It was as simple as that. You could use a Morse key to turn a lightbulb on and off, sending out a visual signal. Ships flashed Morse messages to each other with powerful beams, but could only do this when they were in sight of each other. This was, in a sense, 'wireless' communication. So too were the smoke signals used by Native Americans, or jungle drums, or the simple messages sent across the sea from one island to another by striking a resonant shell with a stick. But to receive any of these messages, you had to be able to see or hear the signals. To send a message over a long distance a relay was needed. Europe had such a system in the early nineteenth century, with 'telegraph' stations positioned on hills. Large wooden arms were moved to relay semaphore signals from one hill to the next. The invention in the 1840s of the electric telegraph, with Morse keys and receivers connected by cables, revolutionised long-distance communication, and the old hilltop telegraph stations fell derelict.

The great potential of the 'Hertzian' waves that Marconi wanted to harness lay in the fact that you did not have to be able to see or hear them to receive them, and you needed no connecting cable to send a signal. How far they could travel through the air Marconi did not know; but that was not the first problem. If you could not hear or see them, how could you detect them? Marconi knew from reading electrical magazines that some ingenious solutions had been found. A French physicist, Edouard Branly, had shown in 1890 that metal filings when scattered in a test tube would not conduct electric current. However, if they were 'hit' by an electric charge the filings clung together, and a current could pass through the tube.

The English Professor Oliver Lodge showed in 1893 that the 'Branly tube' could act as a detector of Hertzian waves. When a spark was generated the invisible electro-magnetic force would, at a distance, cause metal filings to stick together. Lodge called his version of the Branly tube a 'coherer', and showed how it could act as a kind of electronic 'valve'. If the coherer were put into a circuit with wires from each end, the coherer could turn a current on and off. When the filings lay scattered in the tube no current could pass through it. However, when an invisible Hertzian wave hit the tube, the filings instantly clogged together, allowing an electric current to pass through them and the circuit to be closed. It was like a tap that could be turned on or off from a distance. From a few yards away it was possible to send an invisible, inaudible signal from a 'transmitter', which produced Hertzian waves, to a 'receiver', which reacted to them, closing a circuit which might light a bulb or ring a bell.

That was more or less the state of the art when Marconi began his experiments in earnest. What he wanted to be able to do was to activate, at a distance, a Morse printer so that each time he pressed his sending key the signals would show up as dots and dashes on a tape. Batteries powered the printer, and the current from them had to flow through the coherer, which would be 'on' when the filings inside stuck together, and 'off' when they were scattered. It was relatively easy to ring a bell once, but then the metal filings in the coherer stayed stuck together, and the bell would continue to ring even after Marconi had raised the Morse lever and was no longer sending out Hertzian waves. To break the circuit and silence the bell the glass coherer had to be shaken so the metal filings lay scattered once again, and no current could pass through them.

The solution Marconi devised to this problem illustrated his craftsman's genius. Firstly, he experimented for hours to find the best and most sensitive metal filings to put in the coherer. He then made the glass tube smaller and smaller. To do this he used thermometers, which he remoulded using a hand-bellows, heating the glass with a naked flame. He had to create a vacuum inside these miniaturised coherers to increase their efficiency, and tiny silver plugs were used at either end as terminals. Marconi estimated that to make one little coherer took him a thousand hours.

Once he had his super-sensitive mini-coherer working, Marconi devised a little hammer mechanism which was activated each time he raised the lever on his Morse key and cut off the Hertzian waves. The sharp rap the hammer gave to the tiny coherer loosened the metal filings, cutting off the current and silencing the bell. In the same way, it would turn a Morse printer on and off. Hold the key down for a short time, and you produced a dot. Raise the lever, and the printer stopped. Hold the key down again for longer, and you got a dash. It was incredibly slow, but it worked.

It had been relatively easy to make the transmitter. All that was needed was batteries to provide the current, a coil to bump up the charge, and two brass balls fixed so that there was a small gap between them. Press the Morse key and the current flowed; the electricity jumping between the brass balls created a crackling bluish-yellow spark which generated electro-magnetic waves. These waves travelled at the same speed as light – in fact they were a form of light – but the crest between the waves was much longer, and they could not therefore be seen. During thunderstorms lightning gives out Hertzian waves, which is why radios crackle in response to each flash.

Less than a year after the death of Hertz, Marconi had a working wireless system. But if it was to be of any real use, he had to discover if the sparks of his transmitter could send out waves that a receiver could pick up at a distance of more than a few yards. In the searing heat of the summer of 1895 he first took his boxes outside into the parched fields and neatly trimmed vineyards of the Villa Griffone to discover what the limits of his invention were.

There was nothing in any of the electrical magazines he had read which could help him. All he could do was try different arrangements of transmitter and receiver. Possibly recalling Benjamin Franklin's experiment with the kite in a thunderstorm, Marconi had the idea that if he raised a wire in the air and put another in the earth, there might be extra power. He was thrilled to discover that this arrangement worked, and he found that the higher the wire, and the more powerful the spark, the further signals would travel. His brother Alfonso moved the receiver and transmitter further and further apart. When the distance reached about a mile, Alfonso was out of sight on the brow of a hill which rose gently behind the villa, and he or one of the farmhands who was helping had to fire a gun to confirm that a signal had got through. Marconi's father, who had funded his son's madcap experiments with disgruntled reluctance, was at last impressed enough to discuss how this intriguing invention might be turned into a commercial venture.

From the moment he developed his primitive, home-made wireless system, Marconi felt he was in a race against time. If he could achieve the results he had working in the attic and the grounds of the Villa Griffone, surely someone in a university or a telegraphy company would soon come up with the same thing, or something better. If Marconi failed to make his name and his fortune with this invention, he had nothing to fall back on. His father had wanted him to join the Italian navy, but Guglielmo, preoccupied with his experiments, had failed the examinations for entry to the naval college.

Now old Giuseppe accepted that his youngest son's future, if he had one, was with the odd bits of wire and batteries strewn about the attic of the villa, and the strange-looking antennae erected in the grounds of the estate. But who would be interested in Guglielmo's magic boxes? And would anyone invest in them, so that the family fortune would not dwindle away? The Italian ministry of posts? Or the navy, perhaps? According to Marconi family legend, approaches were made, but after a wait of several months they received a polite refusal. This is possible, although no records have been found of any contacts with the government. Perhaps the story was invented later to protect Marconi's reputation as a staunch Italian patriot. As it was, Guglielmo and his mother were soon on their way to London, where there was a much greater chance that his invention might be taken up, with the help of Annie Marconi's many wealthy and influential relatives.

4

※→※←※

In the Heart of the Empire

London, the heart of the British Empire, was a huge metropolis in the last decade of the nineteenth century, with a population of more than six million. The smoke from tens of thousands of chimneys filled the air with the sooty haze that so attracted Impressionist painters such as Claude Monet, who captured on canvas the strange light that hung about the Thames, the great railway stations and the Houses of Parliament. London had had a steam-driven underground railway since the 1860s, and the first of the new electric underground lines had been opened between the suburb of Stockwell and the City in 1890. But the open-topped buses, carriages and goods wagons were still horse-drawn, as were the trams which provided cheaper fares out to the new working-class suburbs. Though gas and petrol engines had been devised, and the first imported motor-car had made a brief appearance on the streets in 1894, the only familiar motorised road transport was steam-driven. These lumbering, steam-engine-like vehicles were kept to a speed limit of four miles per hour in the countryside, and two miles per hour in London and other towns, and were required to have a man walking twenty yards ahead of them; the stipulation that he carry a red flag, first made law in 1865, was dropped in 1878.

Most of the metropolis was still gas-lit, though electric light in one form or another had been around for a number of years. The first experimental electric street lights had been of the carbon arc variety: a fierce, crackling white glow was produced by passing a current through two carbon rods separated by a small gap. These had been used as early as 1878 to floodlight a football match in the northern town of Sheffield, an experiment that was abandoned when the players complained that they were blinded by the glare and could not see the ball. The little town of Godalming in Surrey, to the south of London, had been the first in the world to have public electric street lighting in 1881, but the electric supplier found it uneconomical, and Godalming returned to gas. The façades of one or two London theatres, such as the Gaiety, were brilliantly lit with arc lamps, which were described as like 'half a dozen harvest moons shining at once in the Strand'.

The forerunner of the modern electric lightbulb had been invented simultaneously in the 1870s by Thomas Edison in the United States and Joseph Swan in England, and in 1879 they joined forces as 'Ediswan' and were turning them out in their thousands. But only large institutions and the grander private houses could afford to have a generator installed, whether it was steam-driven or water-powered – the first hydro-electric system was fitted by Edison in Cragside, the stately home of the English arms magnate William Armstrong, in 1880. There were no large power stations in Britain – nothing to compare with the massive turbines driven by Niagara Falls in the United States – and only a handful of people in London could flick a switch to turn on domestic lights. In fact, so unfamiliar were light switches that notices were sometimes placed next to them, warning that no attempt should be made to ignite them with a match.

It was in February 1896 that Guglielmo Marconi and his mother Annie left Bologna and travelled by steam train across Europe, then by ferry to England, arriving at Victoria station in London where the air was thick with the reek of coal-smoke and horse-dung. Henry Jameson-Davis, the son of one of Annie's sisters, who had known Guglielmo as a boy, agreed to find them a place to stay, and was intrigued by his young cousin's wireless equipment. Jameson-Davis was an engineer himself, specialising in the design of windmills, and invited his friends to see Marconi's invention. One of them, A.A. Campbell Swinton, knew William Preece, Chief Electrical Engineer at the Post Office, and agreed to give Marconi a letter of introduction. Dated 30 March 1896, the letter stated:

> I am taking the liberty of sending to you with this note a young Italian of the name of Marconi, who has come over to this country with the idea of getting taken up a new system of telegraphy without wires, at which he has been working. It appears to be based upon the use of Hertzian waves, and Oliver Lodge's coherer, but from what he tells me he appears to have got considerably beyond what I believe other people have done in this line. It has occurred to me that you might possibly be kind enough to see him and hear what he has to say and I also think that what he has done will very likely be of interest to you. Hoping that I am not troubling you too much . . .

In April Marconi wrote home to his father that he had had a meeting with a Mr Price – he got the name wrong – who had shown an interest in wireless. It is not clear exactly when Marconi demonstrated his working model to Preece. A description of Marconi's arrival at the General Post Office building in St Martin's-le-Grand in the City was given years later by a lad who was one of Preece's assistants, P.R. Mullis. While Mullis was going to and fro unloading Preece's brougham, he noticed Marconi examining a little scale model of an ingenious bag-catching device used by the Post Office which enabled trains to take on post without stopping. He recalled that Marconi had with him two large bags. Preece emerged, shook hands with Marconi and polished his gold-rimmed spectacles as the young man unloaded brass knobs, coils and tubes and set them out on the table. Mullis was sent to fetch a Morse key, some batteries and wire. When he returned, Preece looked at his gold pocket-watch, remarked that it was past midday, and told Mullis to take Marconi to the Post Office refreshment bar and to 'see he gets a good dinner on my account'. They were to be back by 2 p.m.

When they returned from lunch, Preece watched as Marconi pressed the Morse key and rang a bell in his receiving apparatus. This greatly intrigued the Post Office chief, who had apparently never witnessed the use of electronic waves in this way. At the end of the day Marconi was invited back, and he and Preece made some adjustments to the equipment with the help of the Post Office workshop. By the end of July Preece felt confident enough to arrange for Marconi to demonstrate his wireless to senior officials of the Post Office. What impressed everyone most was the fact that the signals could be sent three hundred yards, then nearly a mile, and then even further, with that mysterious ability to go straight through stone walls.

Preece had been booked long before to speak on wireless telegraphy at Toynbee Hall in December, and had intended to give an account of his own work sending messages across the sea on the west coast of Scotland. But he now wondered if his system, which required huge lengths of parallel wires to cover quite short distances, was less promising than Marconi's. If it did the same job, Marconi's would certainly be quicker and cheaper to install. Preece decided to take the opportunity of the Toynbee Hall lecture to introduce his protégé to a wider audience, which is how the two of them came to travel to Whitechapel that December evening. It had been as recently as the summer of 1894 that Marconi first conceived the idea of the use of Hertzian waves for telegraphy; in just over two years he was being fêted by the Chief Electrical Engineer of the mighty British Post Office, and by the spring of 1897 he was being pursued by more than one investor interested in his patents.

5

❖❖❖❖❖

Dancing on the Ether

In the springtime migrant birds moving north to their nesting grounds, yellow wagtails and peregrine falcons, cross the Bristol Channel between the northern coasts of the English counties of Devon and Dorset and the south coast of Glamorgan in Wales. They skim over Steepholm and Flatholm Islands, which lie in the middle of the Channel, and rise up over Lavernock Point, a low cliff facing south from the Welsh coast. A colourful host of pink and white flowers attracts the butterflies which dance in the coastal breezes. There are relics of old gun batteries here, which guarded the wide estuary of the Severn River. When William Preece was experimenting with his method of wireless telegraphy in 1892 he chose the three-mile span between Lavernock Point and Flatholm Island, and found that he could establish a link between the two. He had much less success when he tried to create a link to Steepholm Island, which lies more than five miles from Lavernock.

Flatholm to Lavernock seemed to Preece the ideal testing ground for Marconi's novel wireless system, and in May 1897 he arranged for an experiment which would demonstrate the potential of his protégé's equipment. Although Preece had publicly expressed great faith in Marconi's invention, he was not at all sure that these magic boxes could send and receive signals over any great distance. By nature he was a risk-taker, but over his long career some embarrassing experiences had taught him caution. There was the time in

1877 when the young Alexander Graham Bell, not quite thirty years old, was invited to demonstrate his new invention, the telephone, to Queen Victoria at her summer home, Osborne House on the Isle of Wight.

Beginning at 9.30 p.m., Bell, with the help of Preece, had created a link between the main house and a cottage in the grounds so that the Queen could speak to two aristocrats familiar to her. She listened also to a rendition of the song 'Coming Through the Rye' sung by an American journalist called Kate Field, who had been hired to write promotional articles about the telephone. The Queen was most impressed as calls and performances came in from Cowes, Southampton and London. The grand finale was to be 'God Save the Queen', played by a brass band in Southampton. This had been William Preece's patriotic idea. But as Her Majesty waited for the burst of music, the line from Southampton went dead. By the time it was fixed the musicians had packed away their tubas and trumpets and gone home. Not wanting to disappoint everybody after such a successful day, Preece himself put his mouth to the microphone and hummed the national anthem down the line, putting as much oompah into his rendition as he could. After listening for a few moments, Queen Victoria is said to have remarked: 'It is the national anthem, but it is not well played.'

Preece was aware that he might be in for another embarrassing experience in the cause of technological advance when he asked Marconi to send a signal from Flatholm to Lavernock. There was an element of rivalry in the experiment, for Preece regarded himself as an old hand at this business of telegraphing without wires, and wanted a demonstration of the merits of his own system alongside that of the young Italian. Despite his apparent enthusiasm for Marconi's invention, Preece appears to have hedged his bets, and was not convinced that it would have more than a very limited value to the Post Office. It was still a novelty, and might turn out to be no more than that.

While Marconi assembled his transmitter on Flatholm Island and a receiving station on the cliff at Lavernock, Preece had already had lengths of wire put in place, running in parallel on either side of the Bristol Channel. When an electric charge was sent through one of these wires it would emit waves which were picked up by the other, and a charge was made to 'jump' across the space between, just like the 'crossed lines' that had caused problems with telephones and telegraphs in the City. By turning the current on and off, it was possible to send the dots and dashes of Morse code. Preece had used this 'induction' system as a temporary link between the Isle of Skye on the west coast of Scotland and the mainland when the telegraph cable snapped, and knew it worked across the Bristol Channel.

Preece presumably did not imagine he was about to witness anything of historic importance as he waited in the buffeting breezes for Marconi's signals to come through to Lavernock Point. He was, after all, the head of a British government department which dominated cable telegraphy worldwide, and which could hardly be threatened by a young amateur scientist and his makeshift apparatus. Preece had invited a German Professor, Adolphus Slaby of the Technical High School at Charlottenburg, near Berlin, to witness the demonstration. Slaby had read about Marconi's ability to send signals over quite long distances, although he himself had had much less success.

Lavernock Point is about sixty feet above sea level. Marconi had an aerial put up which rose a further sixty feet in the air, and was topped with a zinc cylinder connected to a receiver which had been set up to record signals in Morse code on a tickertape. Another wire went from the receiver down the cliff to the seashore. The transmitter, firing off a crackle of sparks, was three miles away on Flatholm Island. For two tense days nothing came through on the Lavernock receiver. In desperation, Marconi had the receiver taken down to the beach below the cliff, to see if that made any difference. Almost instantly it began to work.

While William Preece did not really grasp the significance of the occasion, an amazed Professor Slaby certainly did. 'It will be for me an ineffaceable recollection,' he said later. 'Five of us stood round the apparatus in a wooden shed as a shelter from the gale, with eyes and ears directed towards the instruments with an attention which was almost painful, and waited for the hoisting of a flag which was the signal that all was ready. Instantaneously we heard the first *tic tac, tic tac*, and saw the Morse instrument print the signals which came to us silently and invisibly from the island rock, whose contour was scarcely visible to the naked eye – came to us dancing on that unknown and mysterious agent the ether!' He wrote up an account of what he had witnessed for the American *Century Magazine*, which was published in April 1898.

> In January, 1897, when the news of Marconi's first successes ran through the newspapers, I myself was earnestly occupied with similar problems. I had not been able to telegraph more than one hundred metres through the air. It was at once clear to me that Marconi must have added something else – something new – to what was already known, whereby he had been able to attain to lengths measured by kilometres. Quickly making up my mind, I travelled to England, where the Bureau of Telegraphs was undertaking experiments on a large scale. Mr. Preece, the celebrated engineer-in-chief of the General Post-Office, in the most courteous and hospitable way, permitted me to take part in these; and in truth what I there saw was something quite new. Marconi had made a discovery. He was working with means the entire meaning of which no one before him had recognised. Only in that way can we explain the secret of his success.

Slaby hurried back to Germany with Marconi's secret, and set to work replicating as best he could the brilliant success of the Bristol Channel experiment.

> Having returned to my home, I went to work at once to repeat the experiments with my own instruments, with the use of Marconi' s wires. Success was instant . . . Meantime

the attention of the German Emperor had been drawn to
the new form of telegraphy ... For carrying out extensive
experiments, the waters of the Havel River near Potsdam
were put at my disposal, as well as the surrounding royal
parks – an actual laboratory of nature under a laughing sky,
in surroundings of paradise! The imperial family delight to
sail and row on the lakes formed by the Havel; therefore
a detachment of sailors is stationed there during the summer, and I was permitted to employ the crews as helpers.

And so it was that Marconi's first benefactor, William Preece,
had unwittingly enabled the nation which was for many years to
be a bitter rival of Britain in the development of wireless telegraphy
to indulge in a blatant piece of industrial espionage. With the
backing of Kaiser Wilhelm II, who wanted Germany to excel in
all fields of technology, and demanded that scientists be given state
backing, Professor Slaby joined forces with others to develop a
Teutonic version of the Italian's new and quite magical means of
communication. Meanwhile, Preece established his own induction
wireless link across the Bristol Channel, and remained sceptical
about the potential of Marconi's use of Hertzian waves.

However, the City of London was mightily impressed. The
potential value of what Marconi had demonstrated out on Salisbury
Plain and at Toynbee Hall, and now across the Bristol Channel,
lay, as far as the City investor was concerned, almost entirely in
the patent rights. If an exclusive legal claim to the mechanism in
the magic boxes could be established, this patent could be sold
around the world, bringing instant riches. As early as March 1896,
barely a month after his arrival in London, Marconi was writing
to his father at the Villa Griffone with details of offers that were
being made to him by various members of the extensive family
contacts of the Jamesons. There was a Mr Wynne, who was related
to his cousins the Robertsons, offering £2400 if Marconi would
allow him to set up a company in which he would be given
half the shares. Then there was another cousin, Ernesto Burn, a

lieutenant in the Royal Engineers, who told Marconi he had a friend who had been paid £10,000 by the British government with a stipend of £2000 a year for 'a discovery useful to the army'.

While Marconi, staying in Bayswater with his mother, conducted a frantic round of meetings in an effort to find a backer for his invention, his father offered advice which reflected his hope that Guglielmo would cash in as fast as he could and return with his spoils to buy a property near the Villa Griffone. Some solace from home arrived in two barrels of Griffone wine, which Marconi arranged to have bottled. In his letters home he pleaded not for wine but for the funds to pay for patent rights not only in England but in Russia, France, Italy, Germany, Austria, Hungary, Spain, India and the United States. In January 1897 he had written to his father:

> I met two American gentlemen who are willing to acquire my patent rights for the United States of America . . . I understand they would give me £10,000 divided as follows: four thousand immediately and six thousand when the patent is granted by the American Government . . . I believe it may be better for me to accept one of these early offers . . . even in the case something goes wrong with the other applications I would still have made a considerable profit.

A sense of urgency, of being on the brink not only of international fame but of a fortune, runs through the letters to old Giuseppe, and was the driving force in Marconi's life after his arrival in London. This conviction that they were onto something which could bring them all riches was evidently shared by his mother's side of the family, and their willingness to gamble a small fortune on Marconi arose from the pressure to prevent others from profiting from his invention. There was, at the same time, an underlying anxiety about the validity of the young man's claim to have devised a genuinely unique technology, for in a very real sense every piece of his ingeniously fashioned and beautifully crafted equipment was derived from the experimental work of others. Marconi himself was acutely aware of this, and it took the very best patent lawyers in London months to find a form of words which amounted to a convincing case that in assembling bits and pieces devised in laboratories in Germany, France, England and Italy – coils, spark gaps and 'coherers' – Guglielmo Marconi had arrived at a unique arrangement.

Marconi's blood had run cold when in 1896 he met on Salisbury Plain a companionable young man, Captain Henry Jackson of the Royal Navy, who told him that he too had been experimenting with Hertzian waves, and had actually built and operated a wireless telegraphy system which had been given a trial run on a battleship, with some success. According to Captain Jackson, as Marconi listened he became crestfallen, and it was only when the naval officer assured him that this work was top secret, and there were no plans to apply for a patent, that he cheered up. William Preece, during his brief honeymoon with Marconi, would insist on basking in the reflected glory of having 'discovered' the Italian inventor, and continued to lecture to audiences around the country on the great value this new sort of wireless telegraphy might have for lightships and lighthouses. Preece's promotion of Marconi infuriated one of the leading English scientists of the day, Professor Oliver Lodge of Liverpool University.

Preece and Lodge had a longstanding feud about the best way to erect lightning conductors – the Post Office had hundreds of them, to protect the telegraphy system from storms – and Lodge could not abide what he regarded as Preece's ill-informed recounting of the miraculous Marconi invention. An undignified spat broke out on the pages of *The Times*. 'It appears that many persons suppose that the method of signalling across space by means of Hertzian waves received by a Branly tube of filings is a new discovery made by Signor Marconi,' Lodge wrote in a letter to *The Times* in June 1897. 'It is well known to physicists, and perhaps the public may be willing to share the information, that I myself showed what was essentially the same plan of signalling in

1894. My apparatus acted vigorously across the college quadrangle, a distance of 60 yards, and I estimated that there would be a response up to a limit of half a mile.'

By that time Marconi had already demonstrated that the range of wireless waves was not as limited as Lodge claimed. Lodge protested that he did not mean that half a mile was the absolute limit, and commended Marconi for working hard 'to develop the method into a commercial success'. In the same letter he continued: 'For all this the full credit is due – I do not suppose that Signor Marconi himself claims any more – but much of the language indulged in during the last few months by writers of popular articles on the subject about "Marconi waves", "important discoveries" and "brilliant novelties" has been more than usually absurd.'

While this storm was brewing between his bearded benefactor and the piqued professor, the Jameson family freed Marconi from Preece's patronage. His father Giuseppe was persuaded to put up the £300 necessary to pay for legal expenses in procuring patents. Then his cousin, the engineer Henry Jameson-Davis, raised £100,000 in the City, mostly from corn merchants connected with the Jameson whiskey business. The Wireless Telegraph and Signal Company was set up with this substantial investment, equivalent to more than £5 million in today's money. It was a commercial venture, the sole purpose of which was to buy the patents and give Marconi the money he needed to continue his experiments. He got sixty thousand of the £1 shares, £15,000 for his patents and £25,000 to spend on research. It was a massive vote of confidence from his mother's family and their business associates.

Henry Jameson-Davis was not acting in a sentimental fashion by raising this huge sum for his cousin. Jameson-Davis was the archetypal Victorian gentleman, a keen foxhunter who would be out with the hounds in Ireland and England as often as six times a week in the winter hunting season. He would not gamble family money on a twenty-three-year-old with an intriguing but largely untried gadget without good reason. He and the other investors hoped to make a fortune when in July 1897 the Wireless Telegraph and Signal Company opened its offices at 28 Mark Lane in the City. By buying the patent rights as soon as they were awarded, the company put William Preece and the British Post Office out of the picture, and left Marconi to get on with the work of demonstrating what a valuable invention the newly formed company owned.

Marconi anticipated that Preece would not take kindly to being supplanted by a family concern, and on 21 July 1897 he wrote to him from the Villa Griffone explaining his position. All the governments of Europe, he said, wanted demonstrations of his equipment, his patents were being disputed by the likes of Professor Oliver Lodge in England and others in America, and he needed money to refine his equipment, take out new patents and fund more ambitious experiments. His letter concluded: 'Hoping that you will continue in your benevolence towards me I beg to state that all your great kindness shall never be forgotten by me in all my life. I shall also do my best to keep the company on amicable terms with the British Government. I hope to be in London on Saturday. Believe me dear Sir, yours truly G. Marconi.'

Naturally enough, Preece replied that the patronage of the British Post Office could no longer be continued. He showed little concern over the loss of control of the new invention, evidently taking the view that it was not going to be of much practical use anyway.

Privately, Preece was pouring cold water on Marconi's spark transmitter in confidential memoranda to the Post Office and the government, suggesting that really there was not much future in it, and in any case the patent was probably not secure, as Oliver Lodge had a prior claim to it. In his Toynbee Hall lecture Preece had said, to the cheers of the audience, that he would see to it that the Post Office would fund Marconi. But the promised £10,000 had not been forthcoming. With his family firm, Marconi now had the funds and the freedom to set up whatever experiments he wished. As he had become convinced that the most promising practical use of wireless was sending messages from ships to shore, he headed for the coast to test the range and flexibility of wireless telegraphy.

6

✦✦✦✦✦

Beside the Seaside

This was the heyday of the English seaside resort, before the new fashion for sunbathing drew the wealthy to the Mediterranean in the summer months. The luxury Blue Trains steamed down to the French Riviera only in winter, when the mild climate attracted the English aristocracy who developed the resorts of Nice and Cannes. Queen Victoria herself liked to stay in Hyères, near Toulon, but not beyond May, when the heat became unbearable and everyone returned north, the French to Honfleur and Deauville on the Normandy coast, and the English to their favoured grand hotels in Eastbourne, Bournemouth and other fashionable seaside towns.

The railways had opened up many resorts to day-trippers from London, and the south coast of England was becoming socially segregated as the 'quality' sought refuge from brash day-tripper resorts like Brighton in the more exclusive havens and coves. Aristocratic and royal families from all over Europe would spend time in Cowes on the Isle of Wight, a short ferry-ride from the coast. In August there was Cowes Regatta, a gathering of the wealthy and the upper crust who raced their huge yachts and enjoyed a splendid social round. Queen Victoria's favourite retreat was Osborne House, close to Cowes, and she spent most summers here in her old age, enjoying the fresh sea air as she was wheeled around the extensive grounds in her bath-chair. On the white chalk clifftops of the island were grand hotels, those on the southern coast with a view across the Channel to France. Among them was the Royal Needles Hotel at Alum Bay, on the very western tip of the island.

It was in rented rooms at the Royal Needles that Marconi established the world's first equipped and functioning wireless telegraphy station in November 1897. An aerial 120 feet high with a wire-netting antenna was erected in the grounds, without, it seems, giving rise to any complaints from other residents. In various rooms of the hotel were pieces of equipment for transmitting and receiving, and workshops where coils of wire were wound, wax was melted for insulation, and metals filed down for experimental versions of the receiver or coherer.

The location was chosen so that Marconi could test his equipment at sea and as a means of communication between ships and the shore. In the summer months, when the coast teemed with tourists and the horse-drawn bathing machines were trundled into the chilly waters of the Channel for women bathers to enjoy a discreet dip, coastal steamers ran regularly from the pier at Alum Bay to the resorts of Bournemouth and Swanage to the west. Marconi negotiated to fit wireless telegraphy equipment to two of these, the *May Flower* and the *Solent*, so that he could test the range and effectiveness of his station at the hotel. When he and his engineers were transmitting, guests were intrigued by the crackle and hiss of the sparks which generated the mysterious and invisible rays that activated the Morse code tickertape on the ships.

English hotels offered the young inventor comfort and fine food, a place where his mother and older brother Alphonso as well as the staff of engineers he was gathering around him could stay. Though Marconi and his mother had no time to enjoy the glamorous social life of London, they were able to find some relaxation on the breezy south coast of England. After Alum Bay another station was opened at the Madeira Hotel in Bournemouth, fifteen miles down the coast. Bournemouth had many distinguished visitors and residents in the nineteenth century: Charles Darwin had stayed there in the 1860s; the beautiful Emilie Charlotte le Breton, known by her stage name Lillie Langtry, lived in a house in Bournemouth provided by her lover, the Prince of Wales, in the 1880s; Robert Louis Stevenson had written *The Strange Case of Dr Jekyll and Mr Hyde* in Bournemouth while recovering from ill-health; and the artist Aubrey Beardsley had only recently left the resort after a period of convalescence when Marconi arrived.

The Bournemouth station first went into operation in January 1898, just before a blizzard blanketed the south coast in deep snow. Newspaper reporters had gathered in Bournemouth, where William Gladstone, the former Prime Minister and grand old man of British politics, was seriously ill. The weight of snow brought down the overhead telegraph wires, and communication with London was cut. It was characteristic of Marconi's opportunism and instinct for publicity that he arranged for the newly opened station at the Madeira Hotel to send wireless messages to the Royal Needles Hotel, from where they could be forwarded to London by the telegraph links from the Isle of Wight, which were still open. In the event, Gladstone recovered sufficiently to return to his home, where he died in May 1898.

Marconi fell out with the management of the Madeira Hotel – it is not clear if the dispute was about money or the nuisance his wireless station caused to other guests – and moved his station to a house in Bournemouth, and then finally to the Haven Hotel, a former coaching inn in the adjoining resort of Poole. This became a home from home for him for many years, long after the station at the Royal Needles Hotel was closed down. After a day in which the Haven's guests were entertained by the crackling of Marconi's aerial, the inventor would often sit down at the piano after supper. Accompanied by his brother Alfonso on the violin and an engineer, Dr Erskine Murray, on cello, the trio would play popular classical pieces as the prevailing south-west wind rattled the hotel windows. Annie Marconi often stayed to look after her son, and those evenings in Poole were among the most delicious and poignant of her life. She was to see Guglielmo less and less as he pursued with steely determination his ambition to transmit wireless signals further and further across the sea. For the time being, however, the fame he had already achieved was a vindication of her faith in him, and indeed of the great risks she had taken in her own life.

7

Texting Queen Victoria

On 8 August 1898 the airwaves crackled with one of the first text messages in history: 'Very anxious to have cricket match between *Crescent* and Royal Yachts Officers. Please ask the Queen whether she would allow match to be played at Osborne. *Crescent* goes to Portsmouth, Monday.' It was sent from the royal yacht *Osborne*, off the Isle of Wight, to a small receiving station set up in a cottage in the grounds of Osborne House. Queen Victoria's reply was tapped back across the sea: 'The Queen approves of the match between the *Crescent* and Royal Yachts Officers being played at Osborne.'

The Queen, then seventy-nine years old, had spent much of the summer at Osborne, and could not fail to notice that something intriguing was going on a few miles to the south at the Royal Needles Hotel. Guglielmo Marconi was not only becoming something of a local figure, he had won tremendous acclaim in the press for one of the first commercial tests of his wireless telegraphy, when the Dublin *Daily Express* had asked him if he could cover the Kingston Regatta in Dublin Bay that July. The newspaper had been impressed by some experiments one of Marconi's engineers had carried out on a treacherous part of the Irish coast for the shipping underwriters Lloyd's of London. To cover the Kingston Regatta Marconi fitted up a tug, the *Flying Huntress*, with his equipment, and followed the yacht races at sea, sending back the latest news and positions to a receiving station on shore which then cabled the up-to-the-minute accounts to the *Express*'s sister paper, the *Evening Mail*.

The *Flying Huntress* was an old puffer, and looked comical with its makeshift aerial mast and a roll of wire rabbit-netting rigged up to exchange signals with the shore station in the gardens of the Kingston habourmaster's home. In contrast to the bizarre sight of 'Marconi's magic netting hanging from an impromptu mast', the Dublin *Daily Express* reporter found the inventor himself captivating.

> A tall, athletic figure, dark hair, steady grey blue eyes, a resolute mouth and an open forehead – such is the young Italian inventor. His manner is at once unassuming to a degree, and yet confident. He speaks freely and fully, and quite frankly defines the limits of his own as of all scientists' knowledge as to the mysterious powers of electricity and ether. At his instrument his face shows a suppressed enthusiasm which is a delightful revelation of character. A youth of twenty-three who can, very literally, evoke spirits from the vasty deep and despatch them on the wings of the wind must naturally feel that he had done something very like picking the lock of Nature's laboratory. Signor Marconi listens to the crack-crack of his instrument with some such wondering interest as Aladdin must have displayed on first hearing the voice of the Genius who had been called up by the friction of his lamp.

There was just as much fascination with the shore station, where Marconi's 'chief assistant' George Kemp, a stocky little Englishman with a handlebar moustache, an indefatigable worker who knew his masts and his ropes from his time in the Navy, and who Marconi had met through Preece at the Post Office, was tracked down by another *Daily Express* journalist. The 'old navy man' gave a down-to-earth account of the state of the art: 'The one thing to do if you expect to find out anything about electricity is to work,' said he, 'for you can do nothing with theories. Signor Marconi's discoveries prove that the professors are all wrong, and now they will have to go and burn their books. Then they will write new ones, which, perhaps some time they will have to burn in their turn.' Of Marconi, Kemp said: 'He works in all weather, and I remember him having to make three attempts to get out past the Needles in a gale before he succeeded. He does not care for storms or rain, but keeps pegging away in the most persistent manner.'

Yet another reporter on board the *Flying Huntress* described Marconi standing by the instruments 'with a certain simple dignity, a quiet pride in his own control of a powerful force, which suggested a great musician conducting the performance of a masterpiece of his own composing'. Though he had been determined not to be overawed by this wonderful invention, the reporter confessed to a thrill when he joined Marconi in a little cabin to send a message to the shore. Having witnessed this remarkable demonstration, a devilish impulse to play with wireless overcame him.

> Is it the Irish characteristic, or is it the common impulse of human nature, that when we find ourselves in command of a great force, by means of which stupendous results can be produced for the benefit of mankind, our first desire is to play tricks with it? No sooner were we alive to the extraordinary fact that it was possible, without connecting wires, to communicate with a station which was miles away and quite invisible to us, than we began to send silly messages, such as to request the man in charge of the Kingston station to be sure to keep sober and not to take too many 'whiskey-and-sodas'.

All the English newspapers reported Marconi's triumph at the Kingston Regatta, and the glowing descriptions of this modest young inventor and his magical abilities impressed Queen Victoria and her eldest son Edward, the Prince of Wales, known affectionately as 'Bertie'.

The Prince of Wales spent much of his time with rich friends, and had been a guest of the Rothschilds, the banking millionaires, in Paris, where he had fallen and seriously injured his leg. In August he was to attend the Cowes Regatta on the royal yacht, and a request was made to Marconi to set up a wireless link between the Queen at Osborne and her son on the ship moored offshore. Marconi was only too happy to oblige: it was excellent publicity, and it was no concern of his if, for the time being, wireless was employed frivolously. In any case, as he later told an audience of professional engineers, it offered him 'the opportunity to study and meditate upon new and interesting elements concerning the influence of hills on wireless communication'.

With an aerial fixed to the mast of the royal yacht and a station set up in a cottage in the grounds of Osborne House, the text-messaging service between the Queen and her son was successfully established. A great many of the guests and members of the royal family on the yacht and staying at Osborne House took the opportunity to make use of this entirely novel means of communication. The messages were received as Morse code printout, which was then decoded and written in longhand on official forms headed 'Naval Telegraphs and Signals'. In this way a lady called Emily Ampthill at Osborne was able to ask a Miss Knollys on the royal yacht: 'Could you come to tea with us some day (end)', to which the reply came: 'Very sorry cannot come to tea. Am leaving Cowes tonight (end)'. More than a hundred messages were sent, many of them from Queen Victoria showing concern for Bertie's bad leg.

This was another triumph for Marconi. He wrote home to his father to tell him excitedly of his two weeks with the world's most famous royal family, that Prince Edward had presented him with a fabulous tiepin, and that he was granted an audience with Queen Victoria. However, what excited him most was the discovery that he could keep in touch with a moving ship up to a distance of fourteen miles, his signals apparently penetrating the cliffs of the Isle of Wight. The newspapers loved it, none more so than a new popular publication which had gone on sale for the first time in 1896, the *Daily Mail*. A full-page illustration showed Marconi at his wireless set, watched by two fascinated ladies, with his signals careering off along a wavy dotted line to the aerial of the royal yacht.

As an inventor Marconi was exceptionally lucky. While others struggled to find financial backers, his contacts through his mother's aristocratic Anglo-Irish family had given him security for at least a year or two, and the money to pay for equipment and assistants. During his brief period under William Preece's patronage Marconi had 'borrowed' the old sailor George Kemp, who became his most loyal attendant. Now Kemp was on the Marconi Wireless Company payroll,* rigging up aerials on windswept coasts wherever they were needed, for all the world like a mariner who had found a new lease of life raising masts on land with which to catch not the wind, but electronic waves. Young as Marconi was, his dedication and single-mindedness, his gentlemanly demeanour, so different from the popular image of the 'mad inventor', and his continuous success inspired loyalty in his small workforce of engineers, most of whom had learned their trade in the business of telegraph cables.

Although a lot of 'secret' experimentation went on in the hotel laboratories on the Isle of Wight and at Poole on the south coast, Marconi was always willing to chance his luck and his reputation with very public demonstrations of wireless telegraphy. This above all endeared him to the new popular journals of the day, which had a hunger for exciting and novel discoveries, especially those which might have potential for driving forward the already wonderful advances in modern civilisation. The dapper figure of Signor Marconi, always smartly dressed, the modest Italian who spoke perfect English and who appeared to be able to work miracles with a few batteries and a baffling array of wires, was irresistible.

* Marconi's name was added to the company name in February 1900. Very rapidly other Marconi companies were formed, including the Marconi International Marine Company and the American Marconi Company, both also in 1900.

his wireless set, a series of two photographs with his signal...
...intentionally done, it was intended that to be the artist...
1962...

8

❖❖❖❖❖❖

An American Investigates

Wherever Marconi went in these heady early days of his fame he was sure to have along with him a writer commissioned by the American *McClure's* magazine. Founded in 1894 by an Irish émigré, Samuel McClure, *McClure's* was one of the first publications to make use of the new process of photo-engraving, which put the old woodcut engravers out of business, as photographs could now be reproduced at a fraction of the cost of hand-carved illustrations. *McClure's* sold for fifteen cents on the news-stands, yet it could attract such eminent writers as Rudyard Kipling and Arthur Conan Doyle. It was the policy of the magazine to invite writers of fiction to cover news events, and *McClure's* fascination with Marconi resulted in a series of wonderfully colourful descriptions of the young inventor at work.

Marconi had already made the headlines with his coverage of the Kingston Regatta and his link-up between Queen Victoria and the Prince of Wales at the Isle of Wight, as well as one or two other well-publicised demonstrations of his invention. When *McClure's* learned that the French government had asked him if he could send a wireless signal across the English Channel – at thirty-two miles by far the greatest distance attempted up to that time – in the spring of 1899, it decided that this had to be covered. Cleveland Moffett, a writer of fictional detective stories, and a fellow reporter, Robert McClure, brother of the magazine's founder, were despatched to cover the historic event, and to reassure themselves and their readers that there was no trickery involved. Moffett joined Marconi on the French side, in the small town of Wimereux, close to Boulogne-sur-Mer, where thirty-five years before Annie Jameson had secretly married Giuseppe Marconi. He wrote:

At five o'clock on the afternoon of Monday, March 27th, everything being ready, Marconi pressed the sending-key for the first cross-channel message. There was nothing different in the transmission from the method grown familiar now through months at the Alum Bay and Poole stations. Transmitter and receiver were quite the same; and a seven-strand copper wire, well insulated and hung from the sprit of a mast 150 feet high, was used. The mast stood in the sand just at sea level, with no height of cliff or bank to give aid.

'Brripp – brripp – brripp – brripp – brrrrrr,' went the transmitter under Marconi's hand. The sparks flashed, and a dozen eyes looked out anxiously upon the sea as it broke fiercely over Napoleon's old fort that rose abandoned in the foreground. Would the message carry all the way to England? Thirty-two miles seemed a long way.

'Brripp – brripp – brrrrr – brripp – brrrrr – brripp – brripp.' So he went, deliberately, with a short message telling them over there that he was using a two-centimeter spark, and signing three V's at the end.

Then he stopped, and the room was silent, with a straining of ears for some sound from the receiver. A moment's pause, and then it came briskly, the usual clicking of dots and dashes as the tape rolled off its message. And there it was, short and commonplace enough, yet vastly important, since it was the first wireless message sent from England to the Continent: First 'V,' the call; then 'M,' meaning, 'Your message is perfect;' then, 'Same here 2 c m s. V V V,' the last being an abbreviation for two centimeters and the conventional finishing signal.

And so, without more ado, the thing was done. The Frenchmen might stare and chatter as they pleased, here was something come into the world to stay. A pronounced success surely, and everybody said so as messages went back and forth, scores of messages, during the following hours and days, and all correct.

For a while the makeshift Wimereux station was besieged by dignitaries of various kinds eager to see this extraordinary invention in action. Among them was a British Army officer, Baden Baden-Powell, brother of Robert Baden-Powell, later the hero of Mafeking and the founder of the Boy Scouts. A particular interest of Baden-Powell was the use of man-lifting kites for reconnaissance in battle, and he was devising models of these which were being tested on Salisbury Plain. Marconi had found them useful for raising a temporary aerial when there was no time to set up a wooden mast, and it was not long before Baden-Powell's patented man-lifting 'Levitor' kites were to prove vital in the development of wireless.

Although he himself was clearly convinced that Marconi was not a charlatan, Cleveland Moffett had been told to double-check that there was no sleight of hand going on with the cross-Channel demonstration. Deceit would not have been all that difficult: there were cables under the sea by which messages could have been passed secretly; or there might have been some pre-arranged set of messages which gave the impression that the sending had been successful. Electricity was exciting, but its properties and potential remained mysterious and magical, and the layman was always in danger of being duped. Moffett continued his account:

> On Wednesday, Mr. Robert McClure and I, by the kindness of Mr. Marconi, were allowed to hold a cross-channel conversation, and, in the interests of our readers, satisfy

ourselves that this wireless telegraphy marvel had really been accomplished. It was about three o'clock when I reached the Boulogne station [actually Wimereux, about three miles from Boulogne]. Mr. Kemp called up the other side thus: 'Moffett arrived. Wishes to send message. Is McClure ready?'

Immediately the receiver clicked off: 'Yes, stand by;' which meant that we must wait for the French officials to talk, since they had the right of way. And talk they did, for a good two hours, keeping the sparks flying and the ether agitated with their messages and inquiries. At last, about five o'clock, I was cheered by this service along the tape: 'If Moffett is there, tell him McClure is ready.' And straightway I handed Mr. Kemp a simple cipher message which I had prepared to test the accuracy of transmission. It ran thus:

MCCLURE, DOVER: Gniteerg morf Ecnarf ot Dnalgne hguorht eht rehte. MOFFETT.

Read on the printed page it is easy to see that this is merely, 'Greeting from France to England through the ether,' each word being spelled backward. For the receiving operator at Dover, however, it was as hopeless a tangle of letters as could have been desired. Therefore was I well pleased when the Boulogne receiver clicked me back the following:

MOFFETT, BOULOGNE: Your message received. It reads all right. Vive Marconi. MCCLURE.

Then I sent this:

MARCONI, DOVER: Hearty congratulations on success of first experiment in sending aerial messages across the English channel. Also best thanks on behalf of editors MCCLURE'S MAGAZINE for assistance in preparation of article. MOFFETT.

And got this reply:

MOFFETT, BOULOGNE: The accurate transmission

of your messages is absolutely convincing. Good-by.
MCCLURE.

Then we clicked back 'Good-by,' and the trial was over.
We were satisfied; yes, more, we were delighted.

9

The Romance of Morse Code

As a boy staying with his cousins in Livorno, Marconi had befriended an elderly blind man, a retired telegraph operator. Marconi would read aloud to him and in return he was taught the Morse code and the technique of tapping it out with a Morse key. This was a skill which had been acquired by thousands of young men, and some women, working in the telegraph business in the latter half of the nineteenth century. Alexander Graham Bell's telephone, which of course required no specialist skills to operate, had not replaced the cable telegraph. Dots and dashes which spelled out letters and punctuation in all languages remained, in Marconi's boyhood and for a very long time afterwards, universal. Morse messages could be sent much greater distances than any phone communication, and they could easily accommodate ciphers, which gave a degree of confidentiality.

It was quite by chance that Morse code proved to be ideally suited to Marconi's primitive spark transmitters, which could only send messages in the form of long and short bursts of electromagnetic waves. In fact, had Morse code not been devised more than half a century before Marconi began to create his wireless system at the Villa Griffone, he would have had to invent something very like it. In all probability, he would not have had the idea of wireless telegraphy at all.

The man who gave his name to the code was Samuel Finley

Breese Morse. He was born in Charlestown, Massachusetts, in 1791 and studied at Yale University, where he took an interest in science. His ambition, however, was to be a great painter. He studied in Europe, and had some success with his landscapes and more dramatic canvases. In London he won prizes for his depiction of *The Dying Hercules* as well as for his sculpture. But in America he found it hard to earn a living. He had an unpaid academic post in New York, and got by painting portraits which fetched only about $15 each. On a trip back from Europe on the sailing packet *Scully* in 1837 he fell into conversation with fellow passengers about the uses of electricity, and conceived the idea of an electric telegraph. It was not an original idea, and Morse was not the man to turn it into a workable invention: he lacked the meticulous craftsmanship which was Marconi's greatest talent. But it was his inspiration which led to the development of the telegraph code which would for ever after bear his name.

Morse's original idea was to assign to hundreds of words a dedicated number, and to use electric current to activate a machine at a distance which would record a series of figures on paper. In September 1837 he set up a demonstration in a lecture theatre at the University of the City of New York, with wires wound around the hall to give a distance of about a third of a mile. It was not a working system, but the prototype for something which with a bit of imagination might be made commercial. Morse and his brother Sidney were the publishers of the *Journal of Commerce*, which was read by, among others, the very inventive Vail family of Speedwell, New Jersey. Stephen Vail, the father, had turned a local blacksmith's works into a thriving iron foundry which had built the steam engine for the SS *Savannah*, which in 1819 had become the first ship to cross the Atlantic powered by paddle-wheels as well as sail. His son Alfred had studied at the University of the City of New York, and saw there by chance one of Morse's telegraph demonstrations. He introduced himself, and with his father's agreement subsequently offered to help develop the system.

Morse had no money, while the Vails had prospered from their steam engines and the casting of hundreds of miles of track for the railways which were beginning to spread out across America. An agreement was signed by which Alfred Vail and his brother George would share with Morse all the rights and rewards of a commercial telegraph system. The Vail brothers went to work on improving the technology, while Morse handled the publicity. A deadline was set for 1 January 1838, by which time Morse wanted to be able to offer the US government and businesses a workable system. It was a tall order: the only available electric cable was milliners' copper wire, used to give a structure to the 'skyscraper' bonnets which were then fashionable. The Vails' first batteries were made of cherry wood, with beeswax as insulation.

The local people in the Vails' town of Speedwell thought Alfred and George had lost their minds as they worked long hours on what was regarded as a crazy venture. Meanwhile, Morse continued to compile his dictionary, assigning to each of five thousand words a specific number – England, for example, was '252'. However, devising a machine which could write '252' proved too much for the Vails. They were close to despair when Alfred had the brainwave that a lever which had an up-and-down movement could more easily mark dots and dashes, and these, rather than whole numbers, could represent letters and numerals. Alfred and George had missed their deadline, but they had cracked the problem.

Alfred feverishly studied the letters of the alphabet, and found that 'E' was used more frequently than any other. He assigned it one dot. Other letters were then given their codes – 'S' became three dots, for example. It took until 1844 before the first commercial telegraphy system using what became known as 'Morse code' went into service in the United States. It was in truth Alfred Vail who devised it, but he allowed Samuel Morse to take the plaudits and enjoy the innumerable international honours which were showered upon him.

Operating Morse keys was an entirely new skill, as was the interpretation of the dots and dashes. With the invention of the telephone receiver a tape printer was no longer necessary, for the operator could simply listen to the urgent staccato of the Morse messages, translating them instantaneously from dots and dashes to letters and words. Very soon those with experience found they could recognise the styles of other individual Morse operators; some claimed they could tell the difference between the styles of men and women. Competitions were held to find the most skilful operators, and, ever attentive to the texture of contemporary life, *McClure's* magazine sent a reporter to one such public demonstration held in New York at the turn of the century. It was a 'fast sending competition', held in a great hall in which 'sets of shining telegraph instruments' had been set up. Most of the audience were themselves telegraph operators, there to see which of the dozen young male contestants was adjudged the best. *McClure's* described the scene:

> One by one the contestants stepped to the test table, and manipulated the key. There was a tense stillness throughout the hall, broken when 'time' was called by a trill of metallic pulsations read by most of the audience as from a printed page. The text of the matter is of no concern, an excerpt from a great speech, a page of blank verse, or only the 'conditions' found at the top of a telegraph form. Speed and accuracy alone are vital. Forty, forty-five, fifty words a minute are rattled off, seven hundred and fifty motions of the wrist and still the limit is not reached. The contestants show the same evidences of strain that characterise the most strenuous physical contest – the dilating nostril, the quick or suspended breathing, the starting eye.
>
> Presently a fair-haired young man takes the chair, self confidence and reserve force in every gesture. Away he goes, and his transmission is as swift and pure as a mountain stream . . . The audience, enthralled, forgets the speed, and hearkens only to the beauty of the sending. On and on fly the dots and dashes, and though it is clear that his

pace is not up to that set by the leaders, nevertheless there
is a finish – an indefinable quality of perfection in the
performance that at the end brings the multitude to its
feet in a spontaneous burst of applause; such an outburst
as might have greeted a great piece of oratory or acting.

Marconi was at this time using exactly the same shining Morse
keys as the contestants used for their New York sending competition. But he could not hope to match their speed, and certainly
nobody would hearken to the beauty of his sending. Each dot and
dash sent by wireless was created by a deafening spark – the operators soon took to wearing earplugs. At the receiving end, if the
operator was using headphones, which was soon the regular practice, the noise of interference was also disturbing and uncomfort-
able. It was like listening intently to a station improperly tuned on
the radio. There was no tuning in the modern sense of finding the
exact point on the wave spectrum to receive a transmitter's signal.
And wireless telegraphy was painfully slow. When in 1897 Marconi
returned briefly to Italy to fulfil his obligation to carry out his
military service, he demonstrated his invention to the navy at La
Spezia. On that occasion each Morse dot required the transmitting
lever to be held down for five seconds, and each dash for fifteen
seconds. The letter 'H' alone (dot, dash, dot, dot, dot) took more
than half a minute to send. The messages relayed to and from the
royal yacht *Osborne* were equally laborious.

Lack of speed was not, however, Marconi's greatest concern. He
needed to prove not only that his system worked, but that it could
span distances which the leading scientists of the day insisted were
unattainable. Unless he could send messages hundreds of miles, he
could never compete with the cable telegraph, and wireless would
have only a restricted value for ships at sea. The widely accepted
view was that because the earth was round, a spark signal transmitted from any point on land would head off over the horizon,
keep going until it reached the upper atmosphere and head out
into space. There was no reason to believe that electro-magnetic

waves would 'hug' the surface of land or sea. However high you raised your transmitting and receiving aerials, the signal could be picked up no further than the line of sight between the topmost points. Marconi, and those working with him, had no theory to contradict the received wisdom. All they could do was carry on blindly, in the hope of demonstrating that the theory was wrong.

The degree to which Marconi was able to inspire confidence in his assistants, most of whom were a good deal older than him – he was just twenty-five in 1899 – was remarkable. Quiet and modest though he was in his dealings with the press, Marconi evidently had a messianic quality in his workshops in the Royal Needles Hotel on the Isle of Wight and the Haven Hotel in Poole. He led by example, often working into the night.

The detective writer turned wireless sleuth for *McClure's*, Cleveland Moffett, visited both these stations in 1899, and chatted to Marconi and the engineers working with him. One of these was Dr Erskine Murray, based at the Haven Hotel, where he sometimes tuned up his cello to make up a trio with Marconi and his brother Alfonso. Moffett wrote:

> ... after a breezy ride across the Channel on the self-reliant side-wheeler 'Lymington', then an hour's railway journey and a carriage jaunt of like duration over gorse-spread sand dunes, I found myself at the Poole Signal Station, really six miles beyond Poole, on a barren promontory. Here the installation is identical with that at the Needles, only on a larger scale, and here two operators are kept busy at experiments, under the direction of Mr. Marconi himself and Dr. Erskine-Murray, one of the company's chief electricians. With the latter I spent two hours in profitable converse.
>
> 'I suppose,' said I, 'this is a fine day for your work?' The sun was shining and the air mild.
>
> 'Not particularly,' said he. 'The fact is, our messages seem to carry best in fog and bad weather. This past winter

we have sent through all kinds of gales and storms without a single breakdown.'

'Don't thunder-storms interfere with you, or electric disturbances?'

'Not in the least.'

'How about the earth's curvature? I suppose that doesn't amount to much just to the Needles?'

'Doesn't it though? Look across, and judge for yourself. It amounts to 100 feet at least. You can only see the head of the Needles lighthouse from here, and that must be 150 feet above the sea. And the big steamers pass there hulls and funnels down.'

'Then the earth's curvature makes no difference with your waves?'

'It has made none up to twenty-five miles, which we have covered from a ship to shore; and in that distance the earth's dip amounts to about 500 feet. If the curvature counted against us then, the messages would have passed some hundreds of feet over the receiving-station; but nothing of the sort happened. So we feel reasonably confident that these Hertzian waves follow around smoothly as the earth curves.'

'And you can send messages through hills, can you not?'

'Easily. We have done so repeatedly.'

'And you can send in all kinds of weather?'

'We can.'

'Then,' said I after some thought, 'if neither land nor sea nor atmospheric conditions can stop you, I don't see why you can't send messages to any distance.'

'So we can,' said the electrician, 'so we can, given a sufficient height of wire. It has become simply a question now of how high a mast you are willing to erect. If you double the height of your mast, you can send a message four times as far. If you treble the height of your mast, you can send a message nine times as far. In other words the law established by our experiments seems to be that the range of distance increases as the square of the mast's height. To start with, you may assume that a wire suspended from an eighty-foot mast will send a message twenty miles. We are doing about that here.'

'Then,' said I, multiplying, 'a mast 160 feet high would send a message eighty miles?'

'Exactly.'

'And a mast 320 feet high would send a message 320 miles;* a mast 640 feet high would send a message 1280 miles; and a mast 1280 feet high would send a message 5120 miles?'

'That's right. So you see if there were another Eiffel Tower in New York, it would be possible to send messages to Paris through the ether and get answers without ocean cables.'

'Do you really think that would be possible?'

'I see no reason to doubt it. What are a few thousand miles to this wonderful ether, which brings us our light every day from millions of miles?'

It was the universal belief among scientists at the time that light waves could not exist in a vacuum, but must travel through *something*; and so too with electro-magnetic waves. What that something was, nobody knew, but it was given the name 'the ether', and was conceived of as a very thin, colourless, odourless jelly, in which the whole of the universe was set. The most common way of explaining 'wireless waves' was with the analogy of throwing a stone into the still waters of a pond and watching the ripples spread out in ever larger circles. There was some popular confusion about the term 'ether', for it was also the name of a gas which was an early form of anaesthetic. When scientists referred to the ether it was sometimes imagined as a vapour, but really it made no differ-

---

* There is an error in the original: this should read '640 miles'.

ence what it was. Marconi believed in its existence, and as far as he was concerned it had some special property which enabled the electro-magnetic waves he sent out to travel much further than they were supposed to.

After the triumph of his Channel transmissions, Marconi's equipment was given a trial by the British Admiralty during naval manoeuvres in the summer of 1899, when the only rival method of long-distance communication was the carrier pigeon. At the same time the huge publicity afforded his experiments attracted the attention of the *New York Herald*'s flamboyant owner Gordon Bennett Jr, son of the famous newspaper proprietor of the same name who had built a reputation on being first with the news. Marconi, still surviving on the investment funds of his City backers and with precious little income, accepted an offer to cover for the *Herald* the America's Cup yacht races, first held in 1851, which would take place in October 1899. When he set out for New York on the Cunard Line's *Aurania* in September it was his first experience of the romance of the Atlantic liners which were to play such an important part in his life.

10

A New York Welcome

In 1899 the night-time spectacle of New York as the great liners were steered on taut hawsers towards their moorings was breathtaking. The Statue of Liberty was floodlit, and the Brooklyn Bridge a blaze of light. While many homes were still lit by gas, the electricity which illuminated public buildings and stores had in just a decade or so turned New York into a dazzling wonder of the modern world. If Guglielmo Marconi was to find real rivals in the exploitation of his new invention it would surely be here, in a land apparently obsessed with electrical power. And yet he was greeted like a conquering hero by American newspapermen when the *Aurania* docked in New York on 21 September. They all wanted to know who this young man was, and were struck by how much he differed from the inventive genius of their lively imaginations. In a report Marconi gave to the Wireless Telegraph and Signal Company back in London he wrote that he had to 'run the gauntlet' of reporters and photographers as soon as he went down the gangway. 'For some reason or other it seemed to come as a shock to the newspapers that I spoke English fluently, in fact "with quite a London accent" as one paper phrased it, and also that I appeared to be very young and did not in the slightest resemble the popular type associated with an inventor in those days in America, that is to say a rather wild haired and eccentrically costumed person.' He did however for a while lose his legendary cool: half his luggage had by mistake been diverted to Boston. In its report of 22 September 1899 the *New York Tribune*, which noted that Marconi explained about his missing luggage in 'good English', commented: 'He is a slight young man of light complexion and nervous temperament, and he is a bit absent-minded. He is evidently more concerned about his scientific studies and inventions than about conventionalities and dress. He has clear blue eyes, and his face is clean shaven, except for a small moustache.' The *Tribune* man had clearly caught Marconi off his guard: it was rare for him to be described as anything but well dressed when he made his public appearances.

No sooner had Marconi and his engineers begun to establish a shore station for the coverage of the America's Cup than his extraordinary celebrity was eclipsed by the return from the Far East of a real American conquering hero. In 1898 the United States had gone to war with Spain. Cuba was then still a Spanish colony, but was being torn apart by a nationalist rebellion which America supported. The United States also coveted the Spanish colonies in the Philippines. When the American battleship *Maine* was blown up in Havana harbour the Spanish were ordered out of Cuba. Under the command of Admiral George Dewey, the American fleet had taken on the Spanish in the Philippines, and at the battle of Manila routed them without losing a single life. In September 1899 Admiral Dewey was on his way back to New York, where he would receive a tumultuous reception. The America's Cup was delayed a few days so that the Governor of New York State, Theodore Roosevelt, could stage the most spectacular 'welcome home' ever witnessed in the United States.

When Dewey and his fleet steamed into New York the Brooklyn Bridge blazed 'WELCOME DEWEY' in lightbulb letters thirty-six feet high and 370 feet across. One thousand bulbs were used for the letter 'W' alone. On Manhattan a victory parade a mile and a half long was lined with wood and plaster statues leading to a Dewey Arch. A two-day holiday was declared, and fireworks cracked in the bright electric air for several nights. Not wanting to be upstaged, the *New York Herald* asked Marconi to throw his wireless equipment on a tug and go out to greet Dewey before he docked. Back in the 1830s James Gordon Bennett Sr had made his name by beating rivals to the European stories arriving on Atlantic ships. As the sailing packets and later the steam liners approached Staten Island, despatch boats were sent out to collect the news and carry it to the *New York Herald* office so that it could be published before the ships docked. Gordon Bennett Jr was the man who sent Henry Stanley to seek out Dr Livingstone in Africa, and Stanley's celebrated formal greeting, 'Dr Livingstone, I presume,' had entered the language. He was also the promoter of a company which was involved in laying the first telegraph cables across the Atlantic. Though the younger Bennett now spent most of his time on his lavish yacht in the Mediterranean, the *Herald* had not lost its competitive edge.

Admiral Dewey arrived in New York two days earlier than expected, and Marconi missed the chance to waylay him at sea. However, the *Herald* was not to be defeated, and was able to arrange for Marconi to take part in a parade of ships held in Admiral Dewey's honour. Shore stations had already been set up on the Navesink Highlands along the New Jersey shoreline and atop a building on 34th Street in New York in preparation for the America's Cup coverage, and Marconi and his engineers set up their equipment on two steamships, the *Ponce* and the *Grande Duchesse*. While they were still working frantically to make contact with the shore stations, the *Herald* had the *Ponce* cruise past Admiral Dewey's flagship, and reported the great cheers that went up when the crowds on both decks were told that Marconi, the wireless genius, was aboard. At one point, according to an ecstatic *Herald* account, a young woman on a ship in the harbour picked up a microphone and called 'Three cheers for Marconi!', to which there was a roared response. Marconi himself, however, stayed at his station, and made no public appearance until he was sure his wireless equipment was working, leaving the captain of the *Ponce* to make his excuses by megaphone.

The *Herald* gave its Marconi wireless coverage a fanfare on Sunday, 1 October 1899, with illustrations of the young Italian's equipment and glowing reports of the way in which he had turned a scientist's dream into an accomplished fact. As he followed the fortunes of the competing yachts, the fabulously wealthy English tea magnate Sir Thomas Lipton's *Shamrock* and the New York Yacht Club's *Columbia*, the news was tapped back to the shore stations on the Navesink Highlands and 34th Street. From there it was sent to Europe and across North America by cable. Marconi worked at first on the *Ponce*, attracting a crowd of passengers who, according to *Herald* reports, were more interested in the inventor than in the progress of the races. There were good commercial reasons for the *Herald* to hire Marconi's wireless telegraphy system, apart from the interest it would attract and the newspapers it would sell. As customers of the various land-line telegraph services, newspapers were always haggling over the cost of using cables for messages. If wireless worked, it would be a serious, and cheaper, rival.

From the day the races began on 3 October 1899 Marconi's fame in the United States was assured. The unbelievable had been achieved. As the *New York Times* put it: 'We at the latter edge of the nineteenth century have become supercilious with regard to the novelties in science; yet our languor may be stirred at the prospect of telegraphing through air and wood and stone without so much as a copper wire to carry the message. We are learning to launch our winged words.' All the newspapers and popular magazines speculated on the future of wireless, the possibility of far-flung families being brought together, of peace descending upon the earth as nation talked to nation with a magic Morse key. And it would all be so much less expensive once the cable companies' monopolies had been destroyed.

In New York Marconi demonstrated his equipment to the United States Navy, which at the time maintained a stock of carrier pigeons for long-distance communication. It went well, but despite enthusiastic reports by their observer on the *Ponce*, the

naval authorities were not sure that wireless was worth the price
Marconi's company was asking, and chose to hedge their bets.
Soon enough, they reasoned, American inventors would come up
with their own version of wireless telegraphy. And in the middle
of October, just as the America's Cup was finishing, a notice
appeared in a number of newspapers in New York to the effect
that Marconi had infringed an American patent taken out as early
as 1882 by a Professor at Tufts University in Boston, Amos Emerson
Dolbear. This patent had been acquired by the Dolbear Electric
Telephone Company of New Jersey in 1886, and then bought by a
Lyman C. Larnard, who was now suing Marconi. Larnard wanted
$100,000 for infringement of his patent, and for all Marconi's dem-
onstrations to be stopped. He told the newspapers that he had bought
Dolbear's patent in July 1899 expressly for the coverage of the
America's Cup, and that he had warned both the *Herald* and Mar-
coni's company that he would sue if they went ahead with their plans.

No notice was taken of this threat, for a brief look at the claim
revealed that what Professor Dolbear had patented was the same
effect of 'induction' that William Preece had used in England.
Lyman C. Larnard had not grasped the difference between this
and the use of Hertzian waves; but then, neither had anyone in
the United States Navy, which would remain woefully ignorant of
wireless technology for almost a decade.

There was for some years a confusion over the difference between
the two methods of 'wireless' telegraphy: Marconi's use of electro-
magnetic waves generated by a spark, and the alternative of 'jumping' currents between parallel wires as employed by Preece and
others. Both worked, and both were genuinely 'wireless'. But there
were two very significant differences. The induction method was
strictly limited in the distance it could cover, as William Preece
had found to his cost. On a Sunday in 1898 he had commandeered
the entire telephone networks down the west coast of England
and the east coast of Ireland in an effort to send Morse signals
across the Irish Sea. All he got was a babble of static; he wondered
if he was picking up unintelligible messages from outer space.

In the United States Thomas Edison had had more success with induction, though over no significant distances. After a poverty-stricken childhood and youth Edison had, through his practical ingenuity, acquired considerable prestige and financial backing, and had set up a powerhouse for electrical experimentation at Menlo Park in New Jersey. While Marconi was still a boy playing with batteries and wires at the Villa Griffone, Edison was demonstrating his brilliantly simple system for sending and receiving telegraph signals from moving trains. All major railroads had running along-side them electric telegraph wires, providing communication between stops along the line. Edison's device involved fitting to the tops of carriages a metal plate which could pick up signals which 'jumped' across the gap of more than twenty feet from the existing wires and transmitted them to a receiver inside the train. Edison had demonstrated this invention in October 1887, on a special train on a section of the Lehigh Valley railroad which ran from New York to Buffalo.

There were 230 distinguished guests aboard, members of the Electric Club and guests of the Consolidated Railway Telegraph Company. As the train flew along, reaching sixty miles an hour at times, four hundred messages were sent. One was relayed direct to London by transatlantic cable. Edison imagined that his invention would be a boon to newspaper reporters and businessmen. How-ever, there was no demand for it: newsmen and businessmen pre-ferred to be free from telegrams of all sorts while 'on the wing'. Marconi's magic boxes were soon to do away with such a leisurely attitude to life.

Edison's induction method worked for a moving train. But it was never going to be any use to a ship at sea, as there were no fixed wires running alongside the liners as they criss-crossed the Atlantic, other than those sunk deep on the ocean bed. Marconi's wireless, however, could be fitted to ships, and in fact to any moving object. Hertzian waves freed wireless to go wherever it was needed, and the means of sending and receiving messages could be packed up neatly in small boxes. That was the beauty of the Marconi system, and in many ways the world of the 1890s appeared to be awaiting his invention.

After a faltering beginning, steamships had conquered the Atlantic, first using the power of paddle-wheels as an aid to sails, then gradually exchanging funnels for masts. New and more efficient engines and the screw propeller cut the crossing times down to five or six days for the swiftest liners, which competed for the right to fly the Blue Riband, awarded to the ship which achieved the fastest crossing. For a very long time after the British government had awarded the Canadian Thomas Cunard the contract to carry mail across the Atlantic in 1838, his shipping line was the leader. Nearly all the large ships were built in Britain, in Belfast or on the Clyde estuary.

In the last twenty years of the century, competition became yet more intense. As they fought to attract the rapidly growing numbers of impoverished Europeans heading for a new life in America, and the wealthy Americans who were beginning to take tours of Europe, the shipping companies ordered larger and more luxurious liners. To have the biggest, fastest ship of the day was good for publicity, even if in other terms it did not make much economic sense. Such was the competition that a new ship was usually out of date within a year or so. Luxury was the keynote of the shipping lines' advertising, which emphasised the romance of shipboard life, often with a wistful illustration of a pretty young lady chatting idly to a handsome officer. The brochures hinted at all kinds of fun – dancing on deck for the steerage passengers, chance meetings of eligible young things in first class. It was a complaint of those who took seafaring seriously that interior designers had taken over the art of shipbuilding, as the staterooms of first class became more and more luxurious.

America, which had had the fastest sailing ships in the early nineteenth century, fell behind in this shipbuilding spree, and the government decreed that only liners built in the United States could fly the Stars and Stripes. This had little effect, but it did produce the liner *St Paul*, which was launched from the

Philadelphia shipyards in 1895, and was to provide the opulent setting for two of the most poignant episodes in Marconi's life.

Once he had satisfied himself that wireless signals could be sent and received over distances which stretched beyond the horizon, and did not disappear into space as the scientists had predicted, Marconi began to dream of conquering the Atlantic. When he first sailed for New York on the *Aurania* it was out of touch with land for days on end. If another ship sailed within signalling distance in mid-ocean they could 'speak' to each other by means of the semaphore flags, but if they hit an iceberg, a common hazard in the North Atlantic in spring, or their engines failed, or they caught fire, they had no means of calling for assistance. Although the Cunard Line had an impeccable safety record, every year passenger and cargo ships disappeared, many leaving no survivors or clues to the fate that had befallen them. When Marconi sailed from New York on 9 November 1899, taking a suite of first-class cabins on the American Line's *St Paul*, he laid plans to end the lonely isolation of ships at sea.

11

$\leftrightarrow\!\!\!\!\leftrightarrow\!\!\!\!\leftrightarrow$

Atlantic Romance

Within the exclusive social circle of first class on the *St Paul*, Marconi was a celebrity, the young inventor all New York had been talking about. But there were those in America who believed that Marconi's fame and popularity were grounded in public ignorance of the new technology. The magazine *Electrical World* saw him off from New York with no more than grudging admiration for his gift for publicity: 'If the visit of Marconi has resulted in no additions to our knowledge of wireless telegraphy, on the other hand, his managers have shown that they have nothing to learn from Yankeedom as to the art of commercial exploitation of an inventor and his inventions.'

Marconi did not, in fact, have any 'managers' orchestrating his publicity, nor did he need any. What had most impressed the newspapers was his refusal to make any claims for his system of wireless that he could not demonstrate publicly. Thomas Edison became one of his greatest admirers, and quipped that the Italian 'delivered more than he promised'. He added that Marconi was the first inventor he had ever met who sported patent leather shoes. In his quiet way, Marconi was an accomplished self-publicist, and before he left New York on the *St Paul* he had devised a scheme which would make the headlines and astonish the first-class passengers on the liner. He arranged for a cable to be sent to the engineers manning the wireless station at the Royal Needles Hotel on the

Isle of Wight, asking them to listen out for a signal from the *St Paul* as it approached the English Channel on the last leg of its voyage to Southampton.

Before he sailed, Marconi set up a wireless cabin on the liner, the first ever on an Atlantic voyage, and tested and tuned it in readiness for the last hours of the journey. The transmitter would have a limited range, of little more than fifty miles, and the *St Paul* would be near the end of its crossing before the Isle of Wight station could pick up its signals.

Before then, Marconi had time to enjoy the easy mid-Atlantic social life. Among the first-class passengers was a glamorous young American woman, Josephine Holman. A family friend of the Holmans', Henry McClure of *McClure's*, a cousin of the magazine's founder, was also aboard, and he no doubt introduced Josephine to Marconi. By the time the *St Paul* was approaching the west coast of Ireland they were engaged. Neither of them was sure how their families would react to the news. Marconi's fame did not necessarily mark him out as a fine 'catch' as far as the parents of marriageable young ladies were concerned. Despite his aristocratic associations through his mother's family, he was fatally Italian, and therefore 'foreign'; and his fortune was by no means assured. The wireless business, many reasoned, might turn out to be just a passing fad. And Marconi's own family might not be keen for him to marry at such an uncertain time in his career, especially to an American woman they had never met. Josephine and Guglielmo decided to keep their engagement secret for the time being.

There was no certainty about when the *St Paul* would enter the English Channel, or when it would be within wireless range of the Isle of Wight. The Marconi engineers waiting at the Royal Needles Hotel were therefore on tenterhooks. In the same way that fishermen attach a bell to their line so that they will know if a fish is biting even if they have dozed off, the engineers had rigged up a system whereby a bell would wake them if their receiver was called up at night. Henry Jameson-Davis and Major Flood Page, the managing director of Marconi's company, were at the hotel awaiting the *St Paul*'s signal. In a letter to *The Times* Major Flood Page gave a vivid description of the excitement of the occasion:

To make assurance doubly sure one of the assistants passed the night in the instrument room, but his night was not disturbed by the ringing of his bell, and we were all left to sleep in peace. Between six and seven a.m. I was down; everything was in order. The Needles resembled pillars of salt as one after the other they were lighted up by the brilliant sunrise. There was a thick haze over the sea, and it would have been possible for the liner to pass the Needles without our catching a sight of her. We chatted away pleasantly with the Haven [the station at the Haven Hotel, in Poole]. Breakfast over, the sun was delicious as we paced on the lawn, but at sea the haze increased to fog; no ordinary signals could have been read from any ship passing the place at which we were.

The idea of failure never entered our minds. So far as we were concerned, we were ready, and we felt complete confidence that the ship would be all right with Mr Marconi himself on board. Yet, as may easily be imagined, we felt in a state of nervous tension. Waiting is ever tedious, but to wait for hours for the first liner that has ever approached these or any other shores with Marconi apparatus on board, and to wait from ten to eleven, when the steamer was expected, onto twelve, to one to two – it was not anxiety, it was certainly not doubt, not lack of confidence, but it was waiting. We sent our signals over and over again, when, in the most natural and ordinary way, our bell rang. It was 4.45 p.m. 'Is that you *St Paul*?' 'Yes.' 'Where are you?' 'Sixty-six nautical miles away.' Need I confess that delight, joy, satisfaction swept away all nervous tension, and in a few minutes we were transcribing, as if it were our daily occupation, four cablegrams for New York, and many telegrams for many parts of England and

France, which had been sent fifty, forty-five, forty miles by 'wireless' to be despatched from the Totland Bay Post Office.

While the rustic Totland Bay post office was handling an unusually heavy load of telegraph messages, including one giving instructions for the menu at a forthcoming dinner party in London, on board the *St Paul* as it steamed towards Southampton there was a good deal of fun and games. The operator at the Royal Needles Hotel tapped out a few bits of news, including the latest from South Africa, where the British were engaged in an embarrassingly costly war with the Boers, who had besieged Ladysmith, Kimberley and Mafeking. With the permission of the ship's commander, Captain Jamison, the on-board printers, accustomed to turning out menus and general notices, produced a small newssheet under the banner *The Transatlantic Times*, vol. 1, no. 1. It was sold to passengers at $1 a copy, the money to go to the Seamen's Fund. One news item bristling with British pluck read: 'At Ladysmith no more killed. Bombardment at Kimberley elicited the destruction of ONE TIN POT. It was auctioned for £200. It is felt that period of anxiety and strain is over, and that our turn has come.'

One of Marconi's engineers, Mr W.W. Bradfield, was credited as 'editor in chief', and Henry McClure as 'managing editor'. Mr Marconi was recorded as having made the arrangements for the publication. And there was a credit for the treasurer – a Miss J.H. Holman. Straitlaced, rather humourless and 'older than his age' in public, Marconi had a keen sense of fun in private. Ever since he had demonstrated his early experiments to his English cousins at the Villa Griffone he had shown a talent for amusing young ladies. He liked to teach them the Morse code, a secret language which in the commercial world was an almost exclusively male preserve, and therefore for young women especially exciting.

Once the *St Paul* had docked, Marconi saw Josephine Holman only occasionally, though she did meet Annie, his mother, in

London. Marconi could not afford much time to enjoy a romance, and Josephine kept in touch with him chiefly by telegraph and letter. She continued to keep their engagement secret after her return to America, fearful that her mother would be furious if she learned of it. When she wrote from her home in Indianapolis she would include passages in the dots and dashes of Morse code, to guard against her mother's watchful eye. In one such passage she described the anxiety she had felt when another suitor had approached her, and her relief when he had proposed to another woman. For his part, Marconi confided in Josephine a secret ambition, and she referred to this darkly in her letters, wishing him luck with the 'great thing' he hoped to achieve.

During the next hectic year, Josephine could often only get news of her fiancé by reading about his endeavours in the newspapers. But these were for the most part downright misleading, as only Josephine Holman and a few other people very close to Marconi knew.

12

❧❧❧❧❧❧

Adventure at Mullion Cove

B lissfully unaware of the epoch-making events taking place a few miles to the south in the autumn of 1899, golfers in plus-fours and tweed caps hacked away on the clifftop fairways of the recently established Mullion Golf Club. The course had been laid out on the windswept western shore of the Lizard peninsula which juts into the English Channel on the southern Cornish coast. Many a ball was lost on the notorious twelfth hole, which was cut through by a ravine between sixty-foot-high cliffs, with Atlantic surf boiling on the rocks below. The club had been established in 1895, when Guglielmo Marconi was experimenting with electro-magnetism in the grounds of the Villa Griffone. In those days the people of Mullion, a pretty fishing village, would still tug a forelock if they encountered Squire Sydney Davey of Bochym Manor. A farm-worker had the task each day of watching the sea from the clifftops for a telltale change in the colour of the water, which would have him running to the village shouting 'Heva, heva, heva!', so the fishermen could get their boats out and net a shoal of pilchards before they swam into the territory of a neighbouring manor. Corn-wall was much more remote than the Isle of Wight or Bourne-mouth, and had had its own language. A donkey pulled the mower which kept the greens trim at the Mullion Golf Club, whose members had reluctantly agreed to pay £6 annually for the rabbit-shooting rights on the course.

This was a deeply romantic corner of England, a treacherous rocky coast with low, crumbling cliffs and sandy coves, where the local people still talked of the lost bounties of wrecked Spanish and Portuguese galleons. Writers and artists sought solace and inspiration in the guest-houses and isolated clifftop hotels which looked out over a sea tossed by winds that blew from the Americas. It was here, close to Mullion, that Arthur Conan Doyle, a visitor to the new golf club, set one of his last Sherlock Holmes stories, 'The Adventure of the Devil's Foot'. In this tale, Holmes had been close to a nervous breakdown, and was persuaded to travel to Cornwall for a rest and a therapeutic dose of sea air. As always, Dr Watson narrated the story, which concerned a strange murder at the vicarage: '. . . we found ourselves together in a small cottage near Poldhu Bay . . . It was a singular spot . . . from the windows of our little white-washed home which stood high upon a grassy headland, we looked down upon a whole sinister semi-circle of Mounts Bay, that old death trap of sailing vessels, with its fringe of black cliffs and surge swept reefs . . . The wise mariner stands far out from that evil place.'

By the summer of 1900, when the clifftops of Mullion and Poldhu Cove shimmered with the pinks and whites of wildflowers, guests staying at the Poldhu Hotel could take the train to Helston station, and then a horse-drawn coach along rough roads up to the clifftop, with fine views to the west over the sea. In August the new arrivals included Major Flood Page, the pipe-smoking and garrulous Chief Engineer Richard Vyvyan, and Guglielmo Marconi. They had no time for golf, for they were here strictly on business, and very soon decided this was just the spot to establish their next, and by far their largest, wireless station. The Poldhu Hotel itself would provide them with a place to stay, food and comfort, but the station they planned could not be accommodated in a few rented hotel rooms. An area next to the hotel atop Angrouse Cliff was leased from its owner, Viscount Clifton, and in October a large area was enclosed with a security fence, and work began on a single-storey building to house the transmitter.

A gate with a stout lock led into the grounds of the Poldhu Hotel.

Surprisingly little notice was taken by the Mullion people of the unfamiliar-looking structure which began to take shape on Angrouse Cliff at the end of 1900. Locals were not unused to industrial developments: Richard Trevithick, the six-foot-two 'Cornish Giant' and famed local wrestler, had designed and run the first ever steam locomotives in the early 1800s. And until the late nineteenth century Cornwall had produced a large part of the world's supply of tin and copper. But this industry had collapsed, and there had been an exodus of Cornish miners to America, South Africa and Australia in the 1890s. Some of those who stayed behind were recruited to work on the wireless station as it took shape early in 1901.

Marconi paid regular visits, but spent most of his time conducting experiments at the Haven Hotel. His mother still fussed about his clothes, even when she was away in London or Ireland or Italy. From Bologna she sent him a letter at about this time: 'I am thinking if it has got warmer at the Haven Hotel you will want your lighter flannels. Mrs Woodward has the keys of your boxes. Your flannels are in the box with the two trays. Summer sleeping-suits on the first tray. Summer shirts under the two trays. Summer suits, jackets, waistcoat and trousers in the wardrobe.'

Early in January 1901 a London newspaper, the *Illustrated Mail*, sent a reporter down to the Haven Hotel to find out what was going on, and published a full-page feature entitled 'A Chat with Mr Marconi'. It began:

> Though bearing a purely Italian name, there is nothing of the foreigner in Mr Marconi's appearance. His speech is that of an educated scientific Englishman and his dress and general manner are those of a pleasant young English gentleman. One understands this when one sees Mr Marconi's mother, who is an English lady, and generally lives with him at Poole harbour. Mr Marconi, of course, finds it necessary to make occasional visits to London, but

with these exceptions and a daily run of an hour or so on his bicycle, all his working hours are devoted to the study of his inventions. Wireless telegraphy has passed out of the sphere of experiment and become an established fact, and Marconi's efforts are now dedicated to the working out of details, the perfecting of the apparatus, and the lengthening of the distance over which electric communication can be effectively worked.

'What distance can you negotiate now?' asked the *Illustrated Mail* representative.

'One hundred miles – or perhaps a bit more – have been successfully worked, and you may take it for granted that in a very short time this distance will be doubled, and even quadrupled,' replied Mr Marconi.

'Is there any truth in reports that you are contemplating the sending of messages between this country and America?'

'Not in the least. I have never suggested such an idea, and though the feat may be accomplished some day, it has as yet hardly been thought of here.'

Marconi was apologetic about the untidiness of the rooms in which he was working, strewn with coils and batteries and lengths of wire: 'This is not by any means a show place. We have no time to spend on beautiful instruments and decorations. We have made hundreds of experiments here, and you will readily understand that there has been no time to waste.' The reporter did not enquire why Marconi was in such a hurry; as far as the readership of the *Illustrated Mail* knew, Marconi had no real rivals. Marconi, however, was acutely aware that there were a great many problems to be solved, and that there were contenders who could overtake him. Though he concealed the real purpose of the Poldhu station from the *Illustrated Mail* reporter, Marconi's ambition was in fact to be the first to send a wireless signal across the Atlantic.

In Germany, Professor Adolphus Slaby had teamed up with

another scientist, Count von Arco, and was publicising his field wireless telegraphy system as a useful addition to army signals. And word was reaching Marconi of a Professor Fessenden in America who had had some success in sending wireless messages over short distances. In Russia there was Alexander Popov, who had built a receiver which picked up at a distance the electro-magnetic waves from lightning bolts, and could forecast the arrival of thunderstorms. Like Oliver Lodge in England, Popov had sent a wireless signal over six hundred yards, for which he had received a Gold Medal at the 1900 Paris Exhibition. In France an instrument-maker, Eugène Ducretet, had built his own version of wireless telegraphy equipment, and as early as 1898 had sent a message from the Eiffel Tower across Paris to the roof of the Pantheon. For all Marconi knew there were others at work who might beat him to the breakthrough he hoped for.

While he concentrated on building a transmitter that would send a signal further than anybody thought possible, Marconi was grappling with another problem: tuning. A wireless signal sent from a spark transmitter could be picked up by anyone who had a receiver. There was no privacy for the sender, no way of narrowing down the signal to a particular wavelength which could only be picked up by a receiver 'tuned in' to it. It was understood that wireless waves varied greatly in length – the distance between the crests of the waves – and that a receiver needed to be tuned to roughly the same wavelength as a transmitter for a signal to be picked up. There was therefore the possibility of dividing up the 'ether' into different wavebands which would not interfere with each other. In 1900 Marconi believed he had got far enough with a solution to this problem to take out a patent which had the number 7777 – known from then on as the 'four sevens patent'. But in reality he and his engineers had only a vague idea what wavelength they were on at any time, for they had no way of measuring the crests between the waves sent out by their transmitters. And even 'tuned' signals were not private, as anyone with adjustable receiving equipment could fish about until they picked up a signal that was not intended for them. This did not matter to passenger liners at sea, which simply wanted to be able to telegraph each other, and harboured no secrets. But for armies and navies, potentially the most important customers for wireless, it was obviously a stumbling block. If the enemy could tap your messages, wireless was likely to be a hazard rather than a help. And even in peacetime there was the problem that if everyone was on more or less the same wavelength, their signals would continually interfere with each other.

For the time being, though, there were few wireless stations in the world, and they were so far from each other that there was no question of eavesdropping or interference. The Marconi companies had most of them in 1900, and they were the only wireless pioneers who had gone in for the manufacture of equipment. The first factory of its kind in the world had been set up in an old silk warehouse in Chelmsford, Essex, to the east of London in 1898, and a workforce, chiefly of women, turned out the spark transmitters and glass tube coherer receivers in anticipation of orders from the British Navy and shipping lines. A huge German liner, the *Kaiser Wilhelm der Grosse*, was the first ship to be fitted with Marconi wireless, W.W. Bradfield, 'editor in chief' of the *Transatlantic Times*, supervising the work. The liner exchanged messages with stations on Borkum Riff lightship and Borkum lighthouse in north Germany.

Two companies with Marconi's name were now in business: the original Wireless Telegraph and Signal Company, which now had 'Marconi' tagged on, and the Marconi International Marine Company, formed with international capital in 1900, but neither was as yet making any money. An infant industry had been founded, and there were just enough customers for wireless to keep it ticking over. Marconi put little store in the achievements he had already made. In 1901 he was determined to put nearly all his energy and the very large sum of £50,000 of company money into his attempt to span the Atlantic. He kept quiet about it because he did not want to damage his credibility by making predictions that would embarrass him if he failed. The distinguished Professor of Electrical Engineering at University College London, Ambrose Fleming, was taken on as a consultant at £500 a year. It was his job to design and test the generator that would be housed in the transmitter building that had been put up next to the Poldhu Hotel in Cornwall. The task of erecting a series of wooden masts around the building to carry a giant spider's web of an aerial was given to the loyal George Kemp, who hired local men and horses for the heavy work that had to be done.

Marconi had been poring over maps of the eastern seaboard of America in search of a place to build a replica of the Poldhu transmitter, and had stuck his pin into Cape Cod. He wanted somewhere which had a clear run of sea to the Poldhu cliff, and the more remote it was, the better: he did not want prying eyes to see the masts going up, or the powerful currents that would be generated to attract attention by interfering with nearby electrical installations. At the beginning of February Marconi left Fleming and Kemp to get on with the Poldhu work, and took with him across the Atlantic Richard Vyvyan, who would supervise the building of the American station. Vyvyan went reluctantly, for he had told Marconi and the company in London that the huge aerials they intended to put up were structurally unsound, and would be vulnerable to high winds. Though Marconi had no engineering expertise, Vyvyan was overruled and ordered to toe the line.

Cape Cod proved to be much wilder than the Lizard peninsula in Cornwall. Marconi did not have the contacts here that he had in England, which made it difficult for him to find a site for his station. In the end, help came from the kind of man he would not normally have had any dealings with, a native of Cape Cod called Ed Cook, who made his living as a 'wrecker'. The busy sea lane into Boston harbour was treacherous and every year ships were driven aground and broken up. On Cape Cod, as elsewhere on the wilder coasts, there were scavengers who searched for cargo washed ashore, and looted the corpses of the drowned. Newspapers still carried shocking reports of wreckers who ignored the cries for help of those drowning just offshore so that there would be plunder for their grim harvest.

The ill-matched pair, Cook in rugged clothing and Marconi in a fur coat – the weather was bitterly cold – explored the coast in a horse and cart. The most suitable site was already occupied by the Highland Light semaphore station, which recorded the passing of ships and cabled news of their arrival to their owners in Boston and New York. The operators working there would not allow Marconi on the site, and he had to settle instead for a clifftop above a bay, where the heavy equipment could be landed by sea and which had a hotel nearby. This was the South Wellfleet, and its hospitality fell well short of that provided by the Haven. After a taste of the local food Marconi refused to try it again, having supplies sent from Boston. He did not stay long at South Wellfleet, and left Richard Vyvyan to organise the delivery of pine posts and all the other equipment needed to build what was then the second-largest wireless station in the world, designed to be almost as powerful as Poldhu. In theory it should be able to send signals the 2300 miles across the Atlantic to Poldhu in Cornwall.

Back at Poldhu, George Kemp was struggling to keep the ring of aerial masts upright as spring and then summer winds frustrated his efforts. Gusts continuously tore away bits of the structure, and Kemp had to search for new sources of timber. On one occasion he went with the coastguards to retrieve the mast of a wrecked ship, and incorporated it into the Poldhu aerial. All the time the generator and power plants devised by Professor Fleming were being tested, sometimes with unexpectedly violent results. Kemp noted in his diary on 9 August 1901: 'We had an electric phenomenon – it was like a terrific clap of thunder over the top of the masts when every stay sparked to earth in spite of the insulated breaks. This caused the horses to stampede and the men to leave the ten acre enclosure in great haste.' For want of any other terminology to describe the unique structure he was working on, Kemp, an ex-naval officer, referred to the various masts and spurs as 'top gallant stays' and 'triatic' stays, for all the world as if he were

Right Marconi, aged about four, with his mother Annie Jameson of the Irish whiskey family and his brother Alfonso at their home, the Villa Griffone near Bologna in Italy.

Below The Villa Griffone shimmering in the summer heat. As a boy Marconi had his home-made laboratory on the top floor, with a window looking onto the rolling vineyards and orchards behind. It was here he first sent wireless signals further than any scientists thought possible.

Previous page Guglielmo Marconi shortly after his arrival in London in 1896. The portrait captures his calm, serious manner, which belied his youth – he was just twenty-two. He described himself as 'an ardent student of electricity'.

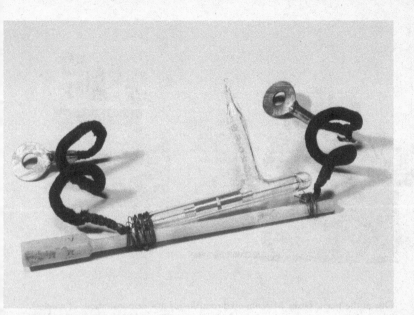

The little 'coherer' which began the radio revolution. Drawing on the discoveries of academic scientists, Marconi crafted this miniature receiver of wireless signals sent in the dots and dashes of Morse code.

William Preece, the distinguished Chief Electrical Engineer of the British Post Office who was Marconi's patron when the young inventor first came to London. After Marconi set up his own company with family help Preece withdrew his support, and later claimed he had invented wireless himself.

One of the 'magic boxes' Marconi used to make his first demonstrations of wireless. This 'coherer' receiver was home-made, but beautifully crafted and 'state of the art' around 1897.

Marconi, fourth from the right, hand on his knee, demonstrates his wireless system to observers from the Italian navy at La Spezia in 1897. Morse code signals are received on the tickertape. Though his invention was funded in London and he spoke perfect English, Marconi remained an Italian patriot all his life.

British Post Office engineers testing Marconi's spark transmitter during experiments in which signals were sent and received across the Bristol Channel in 1897. Though the tests were successful, the Post Office was not as impressed as the German professor Adolphus Slaby, who realised Marconi had discovered something new and dashed home to copy it.

The precipitous location of the world's first working wireless station. In 1897 Marconi rented rooms in the clifftop Royal Needles Hotel on the Isle of Wight and transmitted messages along the coast and to passing steamers taking holidaymakers on day trips along the south coast of England.

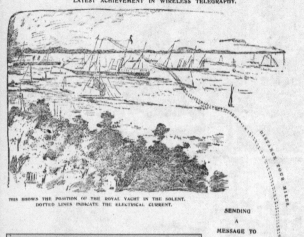

THIS SHOWS THE POSITION OF THE ROYAL YACHT IN THE SOLENT.
DOTTED LINES INDICATE THE ELECTRICAL CURRENT.

SENDING
A
MESSAGE TO
THE
PRINCE.

One of the first wireless 'text messages' in history: it was dictated by Queen Victoria and tapped out by a Marconi engineer in a cottage in the grounds of Osborne House on the Isle of Wight. Marconi himself deciphered the message to Prince Edward on the royal yacht *Osborne*.

The London *Daily Mail*'s graphic illustration on 19 August 1898 of a Marconi engineer tapping out Morse code messages from Lady Battenberg, staying at Osborne on the Isle of Wight, to friends on the royal yacht *Osborne*. Marconi was thrilled to discover that he could keep in touch with his land station while the royal yacht was on the move, and was rewarded with a fabulous tiepin presented to him by Prince Edward.

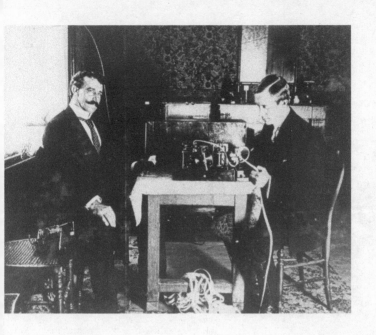

Above The tireless and fiercely loyal ex-naval officer and Post Office employee George Kemp with his boss, the much younger Marconi, in their temporary wireless station set up at Wimereux on the northern coast of France in 1898 to demonstrate the sending of signals across the English Channel.

Left The Haven Hotel, Poole, Dorset, on the south coast of England. Decked out in this picture for the coronation of Edward VII in August 1902, the Haven was Marconi's home and research station for several years.

Above One of Marconi's most talented rivals, the Canadian Reginald Fessenden (centre, with spectacles and watch chain), towering above his assistants at his remote research station at Brant Rock, Massachusetts. It was from here at Christmas 1906 that Fessenden, backed by American financiers, made the first-ever voice broadcasts by wireless. He was disheartened when nobody took any interest in his achievement.

Left Ambrose Fleming, the diminutive and slightly deaf London University Professor who was employed by Marconi's company as a consultant during the crucial years when the first wireless signals were sent across the Atlantic. Fleming's expertise was needed to build the powerful transmitter, and in 1903 he invented the first primitive radio 'valve', adapted from a vacuum lightbulb.

planning to launch this piece of cliff across the Atlantic. Marconi was experimenting with different amounts of power and windings of coil, in search of the most effective arrangement for sending signals long distances. Bit by bit the Poldhu station began to exchange signals with stations sited eastwards along the coast, sometimes with naval stations which had bought Marconi equipment, and sometimes with a Marconi station which had been set up at Crookhaven at the south-western tip of Ireland.

George Kemp had salvaged a small station from the east coast of England and brought it to within a few miles of Poldhu, at the southern tip of the Lizard peninsula. Marconi carried out tests to discover whether the signals from the much more powerful Poldhu transmitter would interfere with those at the Lizard if the two were tuned to different wavelengths. There appeared to be no problem, and while Poldhu was being tested the little Lizard station began to set new distance records, making wireless contact with a second Marconi station on the Isle of Wight at Niton, 186 miles away. Only occasionally did the locals take any interest in what was going on at Poldhu. On 30 May the *Royal Cornwall Gazette* carried a paragraph: 'Mr Marconi accompanied by his private secretary Prof Fleming and several other gentlemen connected with the wireless telegraphy company are staying at Poldhu Hotel, Mullion . . . and from the activity shown it is believed that important experiments are contemplated.' The notion that a distinguished London University Professor was Marconi's 'private secretary' would have caused amusement in the academic world.

If asked, the engineers at Poldhu would say that their 'important experiments' were entirely to do with contacting ships at sea which had been fitted with wireless cabins. The same answer was given to any inquisitive visitors to the station at Cape Cod, and it made perfect sense: there would have to be shore stations on both sides of the Atlantic to make contact with liners if the range of wireless was indeed limited to a couple of hundred miles. It was believed that only very long wireless waves could travel great distances, and that for a signal to cross the Atlantic a charge to the tall aerial towers had to be delivered by a giant version of Marconi's early spark transmitters. By the summer of 1901 Professor Fleming was generating huge sparks at Poldhu that produced thunderclaps which rolled along the Cornish cliffs and echoed in the coves.

After his return from Cape Cod, Marconi spent most of the summer experimenting at Poldhu, while Kemp continued to struggle with the aerial. There was little time for relaxation, and Josephine Holman was beginning to indicate in her letters that her fiancé's long absences and silences were troubling her. An anecdotal story, told years later to Marconi's daughter Degna, has Marconi taking part in a local cricket match at Mullion, perhaps as a member of one of the scratch teams sometimes put together by the guests at the Poldhu Hotel. Sadly the scoresheet is lost, and there is no record of how Marconi, in all other respects the perfect English gentleman, fared with the unfamiliar willow bat and red leather ball.

On Sunday, 15 September George Kemp had a day off, and walked the short distance along the coast to Gunwalloe church for the service. The next day the strong winds and rain made outside work difficult, and Kemp recorded in his diary that he had struggled to keep the masts upright. On Tuesday, 17 September a gale blew in the morning while they were testing different methods of creating sparks. The wind had been from the south-west, but at one o'clock in the afternoon it veered to the north-west, and a sudden squall ripped at the rigging, tore out the supports of one of the masts, and the whole lot collapsed. It was fortunate that nobody was injured. Surveying the wreckage, Marconi asked Kemp to start rebuilding, but to a less complex model. He then took the horse-bus to Helston station and returned to London. He needed to persuade the board members that his plans would have to change. The replacement Poldhu mast would be more robust, but would not have the power of the one that had collapsed. Cape Cod was too far away, and a point on the North American coast closer to Cornwall would have to be found. Wherever that was to be, there would be no time to build a transmitter or even much of a receiving station if Marconi were to achieve his ambition before the end of the year.

To get around the steep, narrow lanes of Cornwall and to speed up his frequent trips to Helston station from the Poldhu Hotel, Marconi had delivered from London a new-fangled machine, a motorbike with an engine mounted over the front wheel. It arrived in kit form, and Kemp helped him assemble it. The sense of urgency at Poldhu was growing every day, as Marconi feared that one morning the newspapers would announce that some other 'wireless wizard' had outdistanced him and stolen his thunder.

13

❧❧❧❦❦❦

An American Forecast

While Marconi was enjoying international fame in 1900, a more seasoned but lesser-known inventor had set up a primitive research station on Cobb Island in the Potomac River, Maryland. Reginald Fessenden, his wife Helen and young son lived a very simple life on the island. With meagre funds, they could not afford the luxurious hotel accommodation Marconi enjoyed. The Fessenden family and their researchers had no running water, and their greatest culinary delicacy was the occasional delivery of oysters brought by a local sea captain. Their remote research station was plagued by insects in summer, which did nothing to soothe Fessenden's notorious fits of temper when his attempts to develop his system of wireless telegraphy were frustrated.

Fessenden was eight years older than Marconi, and a much more experienced experimenter. Born in 1866, the son of a Canadian Anglican minister, he had spent much of his youth close to Niagara Falls at the time when it had first been used for the generation of electricity. He studied mathematics and sciences, but did not finish any course which would have given him a formal qualification. As a teenager he took a job as a teacher in Canada, and when he was only seventeen he moved to Bermuda, where he was the lone teacher in a small educational outfit called the Whitney Institute. While he was there he kept up his interest in electricity, reading the technical magazines. He also met and married his wife Helen.

Determined to play some part in the new electrical industry, he left Bermuda for New York in 1885, and knocked on the door of Thomas Edison. He was turned down several times because of his lack of qualifications, but was eventually taken on to work on the laying of electricity cables, and later as a researcher at Edison's Menlo Park.

Most of Fessenden's work was with chemistry rather than electricity, although the two were closely interconnected. He created a new form of insulating material, and in time became Edison's most senior chemist. He might have gone on to work on Hertzian waves and wireless, but the burgeoning industry of electricity generation for lighting dominated Edison's interest, and the new developments in electro-magnetism were put to one side. In 1890 Fessenden left Edison's General Electric Company and over the next few years moved from one job to another, all the time adding to his inventions and building up his knowledge of electricity generation and its uses. This was an era when the patenting of new devices and inventions had become frenetic in the United States, and Fessenden had many to his name. He also spent some time in England, where he studied British developments and visited James Clerk Maxwell's famous laboratory in Cambridge.

By 1892 Fessenden had a sufficient reputation to be offered the post of Professor of Electrical Engineering at Purdue University in Indiana. It was here that he began to experiment with Hertzian waves. After a year he moved on to Pittsburgh, where he was offered generous grants by Westinghouse, Edison's great rival. When Roentgen's discovery of X-rays was publicised in 1896 Fessenden, like many other scientists, took a great interest in them, and it was some time before he became intrigued by the possibilities of wireless. He claimed it was he who had suggested Marconi to the *New York Herald* as the man to cover the America's Cup in 1899, turning down the job himself. In 1900, however, he accepted an offer from the US Weather Bureau which would enable him to set up a wireless research station, which was how he came to be on Cobb Island.

Though Marconi enjoyed popular acclaim in America, US government departments were reluctant to deal with his company. Britain already dominated the world cable network, and there was a strong desire in America that the same thing should not happen with wireless. Both the Navy and the Weather Bureau wanted to encourage the development of home-grown talent, and Fessenden more or less fitted the bill, though he was Canadian. His task was to create a wireless system which would help the Weather Bureau track and forecast dramatic events such as hurricanes, and he was free to evolve whatever technology he liked, provided it was of some practical value. Marconi's system of spark transmitters and coherer receivers would have done the job reasonably well, with a series of relay stations a hundred miles apart. His great achievement had been to show that wireless waves could travel at least that far, and there was no point in Fessenden repeating that experiment. There was, however, the potential problem of the infringement of Marconi's patents which were being applied for in the United States. Fessenden recognised that, but it did not worry him much. He greatly admired Marconi for his youthful achievements, but felt that the spark–coherer system had done its job, and was in any case fatally flawed: it would never be able to send Morse signals at any great speed, and there was no prospect of it sending and receiving speech.

With scant concern for the expectations of his paymasters, Fessenden turned Cobb Island into something like a pure research station. He devised a new kind of receiver, in which a metal plate was immersed in liquid. This 'barreter', as he called it, was much more sensitive than the coherer, and could receive Morse at much greater speeds. Instead of the intermittent bursts of the spark transmitter, Fessenden wanted one which could generate something like a continuous wave of impulses. He did not care how far his signals would travel: he was intent on quality, not distance. Edison – who called him 'Fessie' – had told him that there was as much chance of transmitting speech as 'jumping over the moon'. But in 1900, with a transmitter that sent out a high-speed stream of sparks and his barreter, Fessenden did manage to send a spoken message to one of his assistants: 'One, two, three, four, is it snowing there, Mr Thiessen? If it is, telegraph me back.' The words were barely audible in the roar of static which the barreter picked up, but it was the first wireless telephone message ever sent.

Unfortunately for Fessenden, neither the US Weather Bureau nor anybody else showed the slightest interest in this achievement, and it was not widely publicised at the time. What possible use could there be for short-distance wireless telephony of such ear-splittingly poor quality when there was already the telephone, and Morse for long-distance communication?

Marconi heard about Fessenden's experiments towards the end of the summer of 1900, but they seemed to pose no immediate threat. Fessenden himself did not make any claims for his break-through, and continued to work for the Weather Bureau, which was sufficiently satisfied with his achievements in telegraphy to give him a new commission. He was to establish three new stations on the eastern seaboard, basing himself on Roanoke Island, North Carolina. This was where the first child of European settlers had been born in the sixteenth century, before the entire pioneer community had disappeared in mysterious circumstances.

Towards the end of 1900 the Cobb Island station was dismantled, loaded onto a schooner and shipped down to North Carolina. Two fifty-foot aerial masts were towed behind the ship, as they could not be loaded onto it. At the helm was the old sea-dog Captain Chiseltine, and among his passengers two of Fessenden's young engineers. As they headed for Roanoke Island they hit rough seas. Fearing for the safety of his ship, the captain began to cut the tow ropes to the wireless masts free. He was restrained by the engineers, and luckily another ship came to the rescue. Fessenden's equipment was saved, and his new station was set up in 1901.

In October that year the stocky, moustachioed figure of George Kemp could be seen working long hours around Mullion Cove. He was still supervising the building of the towers for the new transmitter, ordering cement and arranging for rocks and gravel to be brought up from the beach to provide solid foundations. Almost the only break he got was when he attended the Sunday service at Gunwalloe church. Disturbing news had come from Marconi engineers in America, who had gone to New York to cover the latest America's Cup series, expecting to have the wireless reporting field to themselves as they had the previous year, only to discover that they had a rival. A young American just a few months older than Marconi had put together a working wireless telegraphy system, and managed to find financial backing and a commission from the Publishers' Press Association to cover the races.

Lee de Forest had written to Marconi a year earlier asking for employment, but had received no reply. In the meantime he and some friends had copied Fessenden's barreter receiver and set themselves up in business on very little money. De Forest's ambition was to challenge Marconi, and the America's Cup was the ideal stage on which to demonstrate that the dapper Italian was not the only inventor who could send wireless messages from ship to shore. To the great embarrassment of Marconi's engineers the two systems interfered with each other to such an extent that they had to agree with de Forest to transmit alternately at five-minute intervals.

On Monday, 4 November George Kemp went to Truro to look over some timber. When he returned to the Poldhu Hotel there was a telegram waiting for him which read: 'Please hold yourself in readiness to accompany me to Newfoundland on 16th inst. If you desire holidays you can have them now. Marconi.' The attempt to exchange signals with Cape Cod had finally been abandoned, and Marconi's revised plan was now to be put into action. Newfoundland was five hundred miles nearer to Poldhu than Cape Cod, and there was a better chance that it would be in range of the Poldhu station's new aerial. Kemp took the news that he was to pack his bags for Newfoundland in his stride. The following day he took the train to London, and then to Chelmsford to buy equipment: kites, balloons, hydrogen gas and iron filings, and sulphuric acid for making more Leyden jar batteries if necessary, receiving instruments and aerial wire. He paid a brief visit to his home in London, and was able to give his four children a treat when he took them to the Marconi office in Finch Lane, where they had a fine view of the procession of the Lord Mayor's Show. After four days Kemp joined Marconi and another Marconi company engineer, Percy Paget, in Liverpool, from where they would sail on 26 November.

14

Kite-Flying in Newfoundland

St John's, the chief town of Newfoundland, was wild, isolated and backward. Goats wandered about at will, and children would milk them where they found them. Women collected water in wooden buckets, and men rolled barrels of port down the muddy lanes. There was just one paved street covered in rough cobblestones, the harbour was crammed with schooners, and the whole place reeked of salt cod hung out to dry. Newfoundland had remained a British colony when the Dominion of Canada was created in 1867. Neither Marconi nor Paget had visited this remote outpost of the British Empire before, but George Kemp had seen St John's ablaze a few years earlier, in 1892, when he was in the navy. His ship had gone to the rescue from Halifax, Nova Scotia, and the sailors had helped get the fire under control.

On the evening of 26 November, Marconi, Kemp and Paget boarded the SS *Sardinian* at Liverpool. This was an old ship belonging to the Canadian Allan Line, not one of the floating palaces that had previously taken Marconi across the Atlantic. The Allan Line's main business was carrying immigrants from Europe to Canada, and the passenger lists for the years up to 1901 show many sad cargoes of orphans, waifs and strays shipped by various charities to a new and often harsh life.

Just before the *Sardinian* sailed, the captain handed Marconi a telegram telling him that the Cape Cod wireless aerial had collapsed. One of the masts had narrowly missed Richard Vyvyan, and another had crashed into the instrument building. A few weeks earlier this news would have been disastrous, but now it made no difference to Marconi's plans. When the original Poldhu aerial had blown down he had given up on Cape Cod for the time being. He had got the North American maps out once again, and looked for the nearest piece of shore to Poldhu. This pointed him to exactly the same spot where the first successful submarine cable across the Atlantic had been completed in 1866, laid by Isambard Kingdom Brunel's monster iron ship the *Great Eastern* between the west of Ireland and Newfoundland. It had taken twelve years from the founding of the Anglo-American Cable Company in London in 1854 for the venture to succeed. The *Sardinian*'s course took it at a sedate pace along the same route as the cable-laying ship.

The weather was fine for the first two days, and Marconi and his party could walk on the deck, which was almost deserted as there were few other passengers sailing first class at this time of year. Then, towards the evening on 29 November, the *Sardinian* began to pitch and roll, and the following day they had to shelter, shivering, in the deckhouse as waves broke over the bows and stern of the ship. It was not until 6 December that they saw the harbour of St John's. A thick frost covered the decks and lifeboats of the *Sardinian*. To the north they could see icebergs, and to the south the spouting of whales.

Icebergs were a constant hazard to shipping in these waters. As early as September 1899 a Canadian electrical engineer had written to the editor of the *Halifax Herald* in Nova Scotia, pointing out that with wireless telegraphy ships could send each other details of the latitude and longitude of ice floes; they would also be able to contact the lighthouses that had been built all along that coast. Wireless was regarded as a potential saviour, and Marconi was treated as a celebrity, dining at the Governor's house in St John's. On Sunday, 8 December the devout George Kemp was given a seat in the Governor's pew in St John's Cathedral. He recalled helping to save the blazing building a few years earlier.

In bitter weather, Marconi went in search of a site to raise his aerials and set up his equipment. He chose Signal Hill, where an abandoned military hospital provided shelter and a place to store the kites and balloons. The *Halifax Herald* sent a reporter to watch the wireless wizard at work.

ST. JOHN's, Newfoundland, December 9 – Signor Marconi, the inventor of the wireless telegraph, says he will erect a station on a hill at the entrance of St. John's harbour, and will swing two other wires by means of a small balloon on headlands between here and Cape Race, and by this means will determine the best location for a permanent station with which to communicate with shipping traversing the ocean south of the Grand Banks. He has transmitted messages 225 miles, and expects to reach 400 miles while here.

He believes the weather conditions here are favourable, and if he escapes the heavy breezes which interfere with balloon ascensions, he hopes to complete the work within a month. He must exercise special care in the selection of a permanent station, because some geological formations are more favourable than others for only half that distance. He devotes special attention to connecting with New York liners which run about 140 to 170 miles off Cape Race, believing he will be able to reach them almost in mid-ocean, and so forestall their arrival two and a half days. He will communicate with the Elder-Dempster liner *Lake Champlain* for the Gulf of St. Lawrence, and also with the Cunarders. He is confident that the effect of his work will be to enhance greatly the safety of the Cape Race seaboard, and he has secured the support of the Newfoundland government, which will establish Marconi stations along Labrador next summer.

Marconi told nobody what he, Kemp and Paget were really up to, in case he failed to receive a signal from Poldhu. All three maintained an air of cool, workmanlike detachment, though a long time later Marconi admitted to an interviewer: 'The mere memory of it makes me shudder. It may seem a simple story to the world, but to me it was a question of the life and death of my future.' On 10 December the *Halifax Herald* reporter looked on as Kemp and Paget wrestled with a gas-filled balloon, to which an aerial wire running out of a window in the old fever hospital was attached. The balloon was tethered to the ground by several ropes, but the wind tore it free, and it sailed out of sight across the sea. 'The accident is not uncommon,' the reporter wrote, 'and caused little annoyance.' In reality Marconi and his men were desperate.

All the while, 1800 miles away at Poldhu, at 6.30 a.m. and then again three hours later, sparks a foot long and thick as a man's wrist were being generated in sequences of three short bursts. The ground shook each time the transmitter fired the dots of the letter 'S' in Morse code. At Signal Hill, Marconi, unsure of what wavelength the signal would be on, adjusted his receiver. All that he could hear on his headphones was a wild static picked up by the aerial whenever Kemp, Paget and some local helpers managed to hold a balloon or kite aloft at the appointed times.

On the afternoon of 11 December, while Kemp was wrestling with a hydrogen-filled balloon, he very nearly disappeared into the ether himself. A sudden gust of wind carried off one of the mooring lines, which flew out to sea 'like a shot out of a gun'. Had it been the line Kemp was hanging onto, he would have gone with it. Failure was staring Marconi in the face. His only consolation was that at least he had kept his ambition secret, so the world would not take him for a fool.

Certainly, had Professor Oliver Lodge known what Marconi was up to at Signal Hill, waving man-lifting kites and balloons around in an Arctic gale, he might have dismissed him as a crank. Lodge firmly believed that wireless waves could not cross the Atlantic. However, he was also convinced that it was possible to communicate with the spirits of the dead. In 1901 Lodge spent more time investigating the powers of spiritualism than the potential of his coherer and wireless telegraphy. Scientists believed at the time in the existence of ether, the intangible but magical substance whose remarkable properties were only just being revealed. Whereas the pragmatic Marconi did not care how the ether might spirit a wireless signal across the Atlantic, and ignored scientific theory, Oliver Lodge was content in his certainty that Hertzian waves had very limited scope; he found the possibilities of spiritualism much more exciting.

15

❦❦❦❦

The Spirits of the Ether

At the time Marconi was speculatively flying his balloons and kites above Signal Hill, hoping to hear three dots transmitted from Cornwall, a number of eminent scientists on both sides of the Atlantic were absorbed by research which they thought might bring conclusive proof that there was life after death. The invisible forces that were being revealed by the development of wireless seemed to provide some evidence that the claims of clairvoyants and mystics might, after all, have some foundation. Perhaps individuals with special powers really could act as 'receivers' of invisible and inaudible spiritual signals.

The study of the paranormal was not universally considered at that time to be scientifically disreputable. A 'Society for Psychical Research' had been founded in 1882 by two men who had met at Cambridge University, Edmund Gurney and Frederic Myers. Within two years the SPR, as it was known, had seven hundred members, including sixty academics, many of them from Cambridge, fifty clergymen, members of the armed forces, the poet Alfred Lord Tennyson, Lewis Carroll, eight Fellows of the Royal Society, and scientists with an interest in electro-magnetism.

What was most exciting to the SPR towards the end of 1901 was the belief that one of its founders was attempting to keep in touch with them from beyond the grave. Earlier in the year Frederic Myers had died, and he now seemed to be engaged in a piece of posthumous research of a kind which he could not have carried out when he was alive. Myers, a Professor of Classics, had invented the word 'telepathy' to describe the apparent ability of people to communicate without written or spoken words. In his book *Human Personality and its Survival of Bodily Death*, posthumously published in 1903, he stated that experiments could be made 'from the other side of the gulf, by the efforts of spirits who discern pathways and possibilities which for us are impenetrably dark'. Now psychic mediums in the United States, England, India and other countries were recording strange, incomprehensible texts. They were often taken down by 'automatic writing', where the hand of the medium appeared to be controlled by a spirit force. When the quotations recorded by the international band of spiritual telegraphists were read together they suddenly made sense, and appeared to be the work of the Classical scholar Frederic Myers. Most of the mediums were themselves ignorant of Classical literature, and could not have remembered or made up what they wrote or spoke. The phenomenon became known as 'cross correspondence', and intrigued the SPR for a number of years.

Professor Oliver Lodge had been a close friend of Myers, and for a while after his death had taken over as president of the SPR. Lodge's interest in spiritualism was not purely scientific. It was a woman who had been responsible for encouraging him to pursue his scientific interests, just as it had been for William Preece and Guglielmo Marconi. In his case it was his Aunt Anne, a sister of his mother, a cultured and forceful woman who had persuaded the sixteen-year-old Oliver's parents to let him stay with her in London. She took him to public lectures on scientific and religious matters, which were very popular in the 1860s. Without his aunt's help it is unlikely that Lodge would have been able to break away from the family firm, which sold clay to the Staffordshire potteries, for his father did not regard 'science' as a profession. Aunt Anne had died of cancer, and had told Lodge that if she could she would 'come back'.

In 1889 Mrs Leonore Piper had gained fame in Boston as a

medium, and had convinced some serious thinkers that she was genuine. Professor Myers invited her to England so that he could arrive at his own opinion about her, and Lodge found himself the guinea pig in an experiment to test her abilities. He was astonished by the experience. Through Mrs Piper he heard his Aunt Anne speaking to him in her own 'well remembered voice'. That was proof enough for Oliver Lodge that there was life after death. For him to lend his authority to the belief in spiritualism was not insignificant. By the 1890s he was a well-known personality in the lecture halls, and had a reputation as an academic who took a close interest in social issues and new inventions. He had devised an electro-magnetic method of gathering dust in factories, had demonstrated Edison's primitive phonograph, and regarded Alexander Graham Bell's telephone as the greatest innovation in his lifetime. He was a friend of many of the luminaries of the day, including George Bernard Shaw, who liked to make fun of Lodge's belief in life after death. Shaw wrote to him:

> I do not as a matter of hard critical fact, believe in personal immortality, and I never met anyone who did: a profession in it is always, in my experience, accompanied by conditions which annul it: for, briefly, if what is to survive of me is an angel it will not be me who am not an angel. If I am to leave my headaches and my imbecilities and brutalities behind, I shall leave Bernard Shaw behind, and a good job too.

A month after the announcement by Roentgen of the discovery of X-rays in January 1896, Lodge, who was then the first ever Professor of Physics at Liverpool University, had replicated the equipment and was using it in the treatment of patients. He X-rayed a boy with a bullet in his head, another child with a halfpenny stuck in his throat, and a jumble of strange objects which had become lodged in infant bowels. Lodge's interest in psychic matters should not obscure the fact that he was a practical as well as a theoretical scientist, and a leading light in the exploration of Hertzian waves.

Lodge knew Heinrich Hertz, and believed him to be the true discoverer of wireless. When the young German died in 1894 it was Lodge who gave the memorial lecture at the Royal Institution in London, and demonstrated to the audience the way in which electro-magnetic signals could be sent and received. Four months later he showed the equipment again at a meeting in Oxford, adding a Morse transmitter and receiver provided by Alexander Muirhead, a friend who manufactured telegraph equipment. Another friend of Lodge, the physicist and SPR devotee William Crookes, had predicted with remarkable foresight the development of wireless telegraphy up to the point Marconi had taken it by 1901. As early as 1892 Crookes had written in the *Fortnightly Review*:

> Rays of light will not pierce through a wall, nor, as we know only too well, through a London fog; but electrical vibrations of a yard or more in wave-length will easily pierce such *media*, which to them will be transparent. Here is revealed the bewildering possibility of telegraphy without wires, posts, cables or any of our present day appliances ... an experimentalist at a distance can receive some, if not all, of these rays on a properly constituted instrument, and by concerted signal messages in the Morse code can thus pass from one operator to another.

However, shortly after Hertz died, Professor Lodge began showing more interest in communicating with the dead than fashioning a new technology. In December 1894, while Marconi was working feverishly in the attic of the Villa Griffone with his candles and hand-bellows refining Lodge's coherer, the Professor himself was sitting at a table in semi-darkness as a forty-year-old Italian woman, Eusapia Palladino, used her spiritual power to move chairs and other furniture around the room. Palladino had a 'control', or spirit medium she called St John, who would startle members of her audiences by grabbing their arms. At the session Lodge observed

Mrs Palladino had been hedged about with electrical and other devices in an attempt to detect any cheating, but Lodge came away from it with few doubts.

It was while Lodge was preoccupied with psychic research and the demonstration and practical use of X-rays that he learned that William Preece had taken Marconi under his wing, and was presenting him to learned societies and the general public as some kind of genius. Lodge summed up his feelings about this in a letter to J. Arthur Hill of the SPR: 'Marconi came over with the same thing in a secret box, with aristocratic introductions to Preece of the Government Telegraphs, and was taken up and assisted by him – who was far more ignorant than he ought to have been of what had already been done.'

The truth of the matter was that not only Preece, but William Crookes and Lodge himself should have realised that the invention Marconi was presenting to the world had been discovered nearly twenty years earlier by a brilliant English scientist called David Edward Hughes. Both Preece and William Crookes had witnessed his demonstration of it, without recognising it for what it was. This startling revelation was made in the earliest ever history of wireless, published in 1900, which recalled experiments Hughes had made as far back as 1879. In his *History of Wireless Telegraphy 1838–1899*, the electrician J.J. Fahie described how Hughes had set up a transmitter in his home in Great Portland Street in London, and picked up its signals on a portable receiver which he carried up and down the road outside. Nothing was known then about the existence of 'Hertzian waves', and James Clerk Maxwell had not yet published his theoretical account of how they might behave.

Born in London in 1830, Hughes had been educated at St Joseph's College in Kentucky, after his parents emigrated to America. His interests were music and natural philosophy, but he was also intrigued by the new electric telegraph, and at the age of twenty-six invented a much-improved type-printing machine which was successful in America, but not in England. Hughes took his telegraph printer to Europe, where he made a small fortune from the sale of his invention and received many honours, including the French Légion d'Honneur. For a number of years he lived in Paris, but he was back in London when Bell's telephone appeared in 1876, and set about designing improvements. Within a year he had invented the microphone, and it was while he was experimenting with it that he noticed the effects of small disconnections which appeared to emit signals. He could pick these up with a portable receiver. They seemed to be just clicking sounds with no meaning, but were genuine Hertzian waves, and Hughes showed a number of scientists how he could pick them up with a receiver around Great Portland Street. By the time Marconi arrived in London in 1896 Hughes was a London University Professor, and had long ago abandoned his interest in wireless. As there was no theory to explain what he had discovered it was assumed that he had sent his signals by 'induction', which was not new. Not even William Crookes, who had observed the demonstrations Hughes gave, had guessed that a breakthrough had been made.

For his history of wireless, Fahie tracked Hughes down in 1899, a year before his death at the age of seventy, and persuaded him to give an account of his experiments for the historical record. At the same time he asked for Hughes's opinion of what Marconi had achieved. Hughes told Fahie: 'Marconi has lately demonstrated that by the use of the Hertzian waves and Branly's coherer he has been enabled to transmit and receive aerial waves to a greater distance than previously ever dreamed of by the numerous discoverers and inventors who have worked silently in this field. His efforts at demonstration merit the success he has achieved.'

If anyone had a claim to have invented wireless before Marconi it was David Hughes, and his magnanimous tribute is in sharp contrast to the disgruntled griping of Oliver Lodge. Yet Lodge had not only failed to exploit the technology he claimed to have pioneered, he had no vision at all of its potential. Whereas a lack of scientific theory had undermined the early work of Hughes, by the time Marconi came on the scene a very limited understanding of the behaviour of wireless waves among scientists had discouraged experimentation. What Marconi was undertaking was an act of faith on which the whole of his future depended. On the windswept shores of Newfoundland he was flying kites, figuratively and literally. Had David Hughes lived even another year, he would have been astonished by the claim Marconi was to make in December 1901.

16

❧❧❧❦❦❦

Fishing in the Ether

As blizzards swirled around the makeshift receiving station in the old military hospital at Signal Hill, Marconi sat wrapped in his coat for hours, listening intently for the three Morse code dots fired through the ether from Cornwall. Outside, his men braved the icy winds which blew small icebergs into Glace Bay. When it became clear that the balloons were not going to remain in the air long enough to hold an aerial, Kemp and Paget with some local help began to fly Baden-Powell man-lifting Levitor kites. They were unstable and difficult to handle, but on 12 December, at the appointed hour of 12.30 p.m. St John's time, Marconi thought he picked up the distinctive three dots of the letter 'S' from Poldhu. He had abandoned the Morse printer in favour of an earphone, and had given up on his supposedly 'tuned' circuit. In effect he was fishing desperately in the sky for the telltale signal which would make or break his company. When Marconi thought he had heard the three dots again he handed the phone to his assistant, asking, 'Do you hear anything, Mr Kemp?' Kemp listened intently, and confirmed that he did. Their excitement was intense, but neither leapt in the air with joy. The link with Cornwall could not have been more tenuous, and Marconi was acutely aware that there would be scepticism among other scientists. This was not solely because theory was against him; everyone would know that he was under tremendous commercial pressure to claim success, as this would undoubtedly push up Marconi company shares, and help to recoup some of the £50,000 invested in the project.

With help, Kemp had managed to fly the Levitor kite to a height of over five hundred feet, and in worsening weather he tried many times to repeat what he and Marconi believed was their success in receiving the signal from Poldhu. At one point they took the aerial down the cliff and attached it to an iceberg which had become marooned in the harbour. To Kemp's disappointment they never got to try this aerial, which he felt would 'probably give us better harmony with the earth's electric medium and the transmitter at Poldhu'. Whatever this theory of the propagation of wireless signals was based on, it was soon forgotten, and Marconi frankly admitted that he had no idea how a signal had got from Poldhu to St John's.

Had Marconi been more confident, he would have called in others to witness the reception of the Cornish signals. But he felt that would be too risky. He prepared a telegram to send to the company's London office in Finch Lane: 'Signals are being received. Weather makes continuous tests very difficult. One balloon carried away yesterday.' But he did not dare send it straight away. On Friday the thirteenth he and Kemp satisfied themselves that they had again received the signal, and the following day Marconi sent the telegram to London. He then told the Governor of Newfoundland, Sir Cavendish Boyle, who sent an official telegram to the British government and King Edward VII, who had succeeded his mother Queen Victoria in January that year.

The story broke in the newspapers on Sunday, 15 December, and caused a sensation. That day Marconi, Kemp and Paget went to the Roman Catholic church in St John's and sat close to the organ, next to the choir. The Non-Conformist Kemp noted disapprovingly in his diary that the service was 'more like a concert'. They were invited to lunch by Sir Cavendish, and a bottle of champagne which had been salvaged from a wreck after years lying on the seabed was served, a wrecker's reward for the man whose invention was to be the saviour of ships at sea. Kemp went to the

Presbyterian church in the evening, then stayed up all night writing telegrams.

While Marconi and his assistants were being fêted in St John's, they awaited news of the reaction around the world. Thomas Edison in the United States and Oliver Lodge in England were sceptical about whether a signal from the other side of the Atlantic could actually have been received. Newspapers, on the other hand, were inclined to believe Marconi. He had such a high reputation with journalists, and struck them as so unassuming, that they gave him the benefit of the doubt, and congratulations poured in.

But almost immediately the Anglo-American Cable Company, which was handling all the dots and dashes of adulation arriving in Newfoundland, let Marconi know that they regarded his wireless experiments as in breach of their fifty-year monopoly, which would not expire until 1904. If he did not cease his experiments, they informed him, they would sue. He sent a reply assuring them that his tests would not continue. The newspapers in St John's were furious, accusing the cable company of taking a dog-in-the-manger attitude and holding up the march of scientific exploration. Newfoundland wanted Marconi to stay: he had put it on the world map.

Newspaper reporters began to make their pilgrimage to St John's to seek out the wireless wizard. Gordon Bennett's *New York Herald*, which had first established Marconi's reputation in the United States with the America's Cup coverage, printed a long interview on 16 December headlined 'Mr Marconi Answers his Critics with Details'. The sceptics' view that all he had really heard was 'atmospheric disturbances' was put to him, and he replied that he expected such criticisms, and that many people would find it hard to believe what he claimed. But he protested that he was experienced enough in wireless work to be certain that he had not been mistaken. Thomas Edison was quoted as saying: 'I told the *Herald* last night that I doubted this story and I haven't changed my opinion. I don't believe it. Mr Marconi is a practical business man and is striving to perfect his scheme for wireless maritime telegraphy. I do not believe he has succeeded as yet, and if he had accomplished his object I believe he would make the matter public himself in an authoritative manner, over his own signature.'

Next to the quotation from the doubting Thomas Edison, the *Herald* printed a bordered box with a picture of a moustachioed Marconi and the cross-head 'Signals an "Absolute Fact" Declares Mr Marconi'. There was a short statement with a verification by the reporter in St John's that this confirmation was from the inventor himself. That was sufficient to convince Edison, who thereafter gave Marconi his invaluable endorsement.

With the scientific community still divided in their opinion, Marconi's plan was to travel to Cape Cod to see how his American station was progressing, and then to England. The Anglo-American Cable Company threat was sufficient reason to abandon St John's and, as it turned out, a convenient excuse to postpone any plans for setting up a permanent station there.

However, just as they were packing up the kites and balloons and all their makeshift paraphernalia, Marconi was approached by a Canadian who happened to be in St John's at the time of the great excitement of 15 December. William Smith, the Secretary of the Canadian Post Office, suggested that Marconi stay on for a few days, as he might be able to persuade his government that it would be to their advantage to offer Marconi a site for a wireless station just south of Newfoundland, on the island of Cape Breton. This was in Canada, where the Anglo-American Cable Company's monopoly on telegraphy had no force. After an exchange of cable messages with the company in London, Marconi decided to wait a few more days in St John's.

Three days later a telegram arrived from Ottawa offering enthusiastic Canadian government support. This was a piece of luck neither Marconi nor his hardheaded business associates in London could ever have anticipated: the government of a country with enormous shipping interests was offering not only land on which to set up a wireless station, but the funds with which to build it. Marconi accepted an invitation to go to North Sydney on

Cape Breton Island to meet his potential backers. On Christmas Eve he and Kemp took a sleigh ride around St John's to say farewell to all those who had been so hospitable to them, and who were now extremely unhappy that the cable company had driven the inventor away from them and into the hands of the Canadians. Mr Paget packed up what was left of their equipment and had it loaded onto the SS *Sardinian*, which was ready to ship it back to England.

With Marconi and Kemp as they prepared to leave for Nova Scotia on Christmas Day was yet another attentive representative of *McClure's* magazine, this time a writer called Ray Standard Baker who would travel with them on the short sea journey to Cape Breton. A private car in the train called the *Terra Nova* was put at their disposal to take them from St John's to the steamer *Bruce* at Port-aux-Basques. Baker wrote: 'The people of the "ancient colony" of Newfoundland, famed for their hospitality, crowned him with every honor in their power . . . it seemed as if every fisher and farmer in that wild country had heard of him, for when the train stopped they came crowding to look in at the window.'

The owner of the train made sure they were well looked-after, serving champagne, turkey and plum pudding as a blizzard raged around the train. Baker saw Marconi drop his guard just once: 'The only elation I saw him express was over the attack of the cable monopoly in Newfoundland, which he regarded as the greatest tribute that could have been paid to his achievement. During all his life, opposition has been his keenest spur to greater effort.' Not only had the Anglo-American Cable Company driven the inventor into the arms of friendly Canadians, its action was likely to convince the general public more than the pronouncements of any scientist that wireless telegraphy was well on the way to becoming a commercial reality, and a rival to cable.

The shares of the cable companies had already begun to wobble on the stock exchanges. As P.T. McGrath, editor of the *Evening Herald* in St John's, explained to *The Century Illustrated*, a monthly magazine: 'A transatlantic cable represents an initial outlay of at least three million dollars. A Marconi station can be built for only $60,000 ... There are now fourteen cables of various sizes [across the Atlantic] with a total length of 189,000 nautical miles ... these require a great number of ocean going cable steamers for their laying and repairs ...' Wireless, if it could be made to work, was a lot cheaper than cables, which were not only costly to lay, but had to be constantly maintained by fleets of ships which dragged them from the ocean bed for repairs.

After a wild crossing, Marconi and Kemp arrived in North Sydney, Nova Scotia, on 26 December, eleven days after Marconi had announced his Atlantic triumph. He was met by a most distinguished gathering, including the Premier of Nova Scotia, George Murray. At the Belmont Hotel Marconi gave an impromptu press conference, in which he laid before a rapt audience of local journalists and dignitaries a vision of a huge wireless station somewhere on the coast of Cape Breton which would provide a transatlantic service at much cheaper rates than the cable companies could offer.

An electric tram took them to Sydney, where prominent businessmen and politicians were anxious to assure the young inventor that he would have no harassment from any rival organisations in Nova Scotia, and that everything would be done to find the very best site for his station on the rugged coastline. The Dominion Steel and Coal Company gave Marconi a free hand to look for a site on land it owned, and a tour party set off along the coast in a small tug. In blizzards and high seas the boat very nearly collided with another sent out from the port of Louisberg to greet them with the Mayor and most of the town council aboard. Back in Sydney a banquet was held in Marconi's honour, and all Cape Breton was agog to know which district he would choose for his station. On 30 December the *Halifax Herald* exclaimed that Cape Breton would be 'a part of the great theatre in which this masterpiece will be shown to the world by the Anglo-Italian inventor'.

The snow was knee-deep in Nova Scotia, and Kemp jotted in his diary that they were glad to get away from it as they travelled by boat and train to the Canadian capital of Ottawa, arriving on

30 December. It was worth the journey. As well as wining and dining Marconi, the government offered $80,000 towards the cost of building and fitting out a wireless station at Cape Breton. In return they wanted a guaranteed cheap rate on wireless telegrams, undercutting the cable companies by more than 50 per cent, and a fixed rate for ship to shore messages of five cents a word. With a draft agreement drawn up, Marconi and Kemp continued on their lap of honour in North America. On 10 January 1902 they were in Montreal as the guests of the Association of Electrical Engineers of Canada, and that evening they took the night train down to New York through a beautiful arctic landscape which they could survey from their private stateroom. The following morning they saw the Hudson River, frozen solid and covered with a new fall of snow.

Back in Halifax, the *Herald* newspaper was waxing lyrical, calling Marconi 'the most famous young man in the world'. Who is he, the newspaper asked?

> In manner he is modesty itself. Marconi is not one of those men who forever are obtruding their theories or their personality upon one. His disposition is positively retiring. Yet there is in his expressive bright eye an indication of thorough determination – there is that in his whole makeup. Not infrequently a far away look comes over his face, telling almost as plainly as if spoken in words, that his thoughts are following the ethereal wave that carries his signals from point to point across the sea, or that conveys it from 'ships that pass in the night' to far-off unseen shores. He has that about his countenance which one would expect to see in a man possessed of a great idea, but yet there is not the slightest trace of self-engrossment. Marconi seems to be a very 'human' young man, courteous, polite and considerate; he is frank and open, believing that straight-forward dealing is best for himself and best for the cause he has at heart; he appears to take the world

into his confidence, but, of course, at the same time, he knows there are times when it is best not to speak.

One of those times – which had lasted a full year – was when he was attempting to realise his dream of sending a signal across the Atlantic, which he kept from the public until it was fulfilled.

17

The End of the Affair

Late in the evening on 12 January 1902 Guglielmo Marconi and George Kemp arrived at Grand Central Station, New York, and took a horse-drawn cab to their hotel. At the dawn of the twentieth century New York was undoubtedly the most exciting city in the world, with the first of the skyscrapers pushing up to form Manhattan's celebrated silhouette. The excavations for the subway had begun, and motor-cars, many of them powered by electricity, jostled among the wagons on the streets. There was a craze for what was known as 'celestial advertising', the projection of images onto low clouds with powerful carbon arc lamps. On the roof of Joseph Pulitzer's *World* skyscraper was a huge electrically-lit magic lantern weighing over three thousand pounds and casting beams of 1.5 million candlepower. When the nights were clear and there was no cloud cover to act as a screen, images and information were projected onto nearby buildings. In 1891 the *World*'s lantern was used to flash Morse messages on the clouds, giving the latest results in the elections for governor. These could be read in the sky as far away as New Jersey and Long Island. For the presidential elections of 1892 the *Herald*'s searchlight was in Madison Square Gardens; if it shone southwards it meant New York had gone for the Democrat Grover Cleveland, whereas if the city had gone for his Republican rival Benjamin Harrison the beam would have shone to Harlem.

One or two scientists had suggested in the 1890s that light could be used to send global or even extra-terrestrial messages. In a journal called *Science Siftings* there was a quite serious proposal to send messages from New York to London using a huge reflector to bounce the sun's rays off the moon, the short and long flashes spelling out a Morse message in the sky. Professor Dolbear of Tufts University, who had patented the wireless induction apparatus which he claimed in 1899 that Marconi had stolen, suggested that communication with Mars would be possible with a beam of a few million candlepower – but only if that distant planet were inhabited by people as sophisticated as Americans.

The eccentric Nikola Tesla, famous in the United States for his work on the electricity-generating plant at Niagara Falls, was confidently predicting that experiments he had carried out in Colorado, in which huge amounts of electric current had been created, would enable him to recharge power stations from a distance, without any wires. In an article in *Collier's Weekly* in February 1901 he claimed to have invented his own system of telegraphy without wires as early as 1893, and declared that 'the time is not far away now when the practical results of my labors will be placed before the world and their influence felt everywhere. One of the immediate consequences will be the transmission of messages without wires, over sea or land, to an immense distance. I have already demonstrated, by crucial tests, the practicability of signalling by my system from one to any other point of the globe, no matter how remote, and I shall soon convert the disbelievers.' Tesla was a rather unstable character, prone to fits of depression followed by wild elation. His lectures were spectacular. On stage he would pass huge currents of hundreds of thousands of volts through his body, so that white flames shot from his fingers. In darkened rooms he would generate voltages so high they produced a mysterious electrostatic hum.

The wonders of electricity were not yet being enjoyed in the average American home, although the Electric Girl Lighting Company offered glowing waitresses and hostesses for special occasions. Customers could choose from a variety of decorative filament lamp adornments, each of the girls being guaranteed at 'fifty candle-power'. Thomas Edison was keen on this sort of thing, and one of his engineers fitted out his own daughter for a New Year's Eve party with an electric wand, earrings and breastpin, her hair glowing with tiny Edison bulbs.

The American Institute of Electrical Engineers had been due to hold its annual dinner at the Waldorf-Astoria Hotel shortly after Marconi's announcement that he had sent a wireless signal across the Atlantic. There was at first considerable doubt about the claim, but Marconi's modest manner and reputation for caution in all he said won the day, and the Institute had decided to invite him to be guest of honour at the dinner, which was postponed until 13 January 1902, the day after Marconi and Kemp arrived in New York.

The scene that greeted them when they entered the crowded Astor Gallery of the Waldorf-Astoria took their breath away. Behind the guests' table was a black tablet decorated with smilax, a kind of briar, and set with lamps which spelled the name 'MAR-CONI'. At the eastern end of the gallery was another tablet with 'POLDHU' written in lightbulbs, and at the western end another tablet which glowed 'ST JOHN'S'. Between the two tablets was a silk cord with, at intervals, the three dots of the Morse letter 'S' in lights, which appeared to travel from one to the other. There were lights all over the tables, which were decorated with more smilax and American Beauty roses. Among the guests were the British and Italian consuls. The *New York Times* reported:

> The signal for the first applause of the evening was the entrance of a long procession of waiters bearing aloft the ices, which were surmounted by telegraphy poles, steam-ships and sailing vessels fitted with wireless signalling apparatus. The telegraph poles were made of solid ice. 'Frozen out' was the prophetic cry of the diners as they saw them. Marconi arose and clapped his hands in glee when he observed the beautiful procession. Then the signal 'S' began to flash from Poldhu to St John's and back again.

Guests mobbed Marconi, asking him to sign the specially designed menu cards, which had a cameo photograph of the inventor with the three dots of the Morse 'S' arching across from one lighthouse representing Poldhu to another representing St John's. The President of the Institute, Charles Proteus S. Steinmetz, had to call for order so that apologies for absences could be read out. There was 'prolonged and loud applause', said the *Times* report, when the toastmaster read a telegram from Thomas Edison: 'I am sorry not to be present to pay my respects to Mr Marconi. I would like to meet that young man who has had the monumental audacity to attempt and succeed in jumping an electric wave clear across the Atlantic.'

The toastmaster then told the audience that within the last ten days he had spoken to Edison, who had told him he thought that wireless waves must at some time be sent across the Atlantic, but that he had been so preoccupied he did not have time to do it himself. Amidst roars of laughter he continued with a quotation from Edison: 'I'm glad he did it. That fellow's work puts him in the same class as me. It's a good thing we caught him young!' There was a message too from Nikola Tesla, received with loud cheers: 'Marconi is a splendid worker and a deep thinker, and may he prove one of those whose powers increase and reach out for the good of the race and the honour of his country.'

Many more letters of congratulation were read out to loud cheering before Marconi himself rose to make his speech as guest of honour with, as the *New York Times* put it, 'a modesty almost amounting to diffidence'. He spoke of the achievements of his 'system' so far – seventy ships equipped with wireless, the majority on British and Italian naval vessels, the rest on passenger liners, and twenty shore stations in England. His explanation of why he had gone to Newfoundland rather than make the first experiment in transatlantic signalling to the United States was that he thought it 'prudent' to attempt a shorter distance with temporary equipment, rather than wait until his large stations were operating. The implication was that he had not himself been sure that it was possible to cover the distance, and wanted some proof before pursuing the costly work on Cape Cod. He described the trouble he and his team had had with kites and balloons, and was cheered when he described the final triumph on 12 December. Bad weather and the writ from the Anglo-American Cable Company had put a premature end to the experiment, though the manager of Anglo-American in St John's had told Marconi that his achievement had been good for business: in the three days of the experiments fifty thousand words had been sent by cable. And he paid tribute to Alexander Graham Bell, who was in the audience: he had used a telephone handset to receive the signals in St John's.

The support of the Canadian government had clearly boosted Marconi's confidence, and he told the audience that wireless telegraphy across the Atlantic would be much cheaper than cable, making it affordable to people of modest means. Thanking America for its hospitality, he then took his audience by surprise with a toast to the Institute. The *New York Times* reporter wrote: 'Mr Marconi took a glass from the table, holding it high above his head, lowered it to his lips, and started to drink before the diners grasped the situation. Then the men and women quickly found glasses and drank in silence the toast. In a few seconds cheers and hand clapping resounded through the banquet hall while Mr Marconi bowed acknowledgement to the plaudits.'

Just below the report on Marconi's triumph was a short paragraph on 'persistent rumors' that the wireless inventor would be married at the Waldorf-Astoria that week. 'Prof Marconi' – he had been awarded instant academic status – denied the reports, and said he would not be married until his return from Europe in about three months' time. But exactly what had happened between him and the American heiress he had met on the liner *St Paul* was a mystery. The newspapermen wanted to know, none more so than those far away in St John's, Newfoundland, who had taken Marconi to their hearts. A despatch in the *Daily News*, St John's, dated 22 January, headlined 'Marconi Free: Love's Young Dream', gave the news that Josephine Holman's family had announced the day before that she had officially broken off the engagement, which had been made public on 27 April 1901. This was a great surprise, for only a month earlier Josephine had told a reporter:

> I would rather marry that kind of man than a king. I was the first person in the whole world to know of his great plans for long distance telegraphy and I was one of the first to learn of his grand triumph in receiving a wireless message across the Atlantic. It has been a terrible state secret with me for more than a year. On the ocean where we first met he confided in me his hopes and expectations under a pledge of secrecy and we decided we would not be married until the outcome was known. I am proud of his triumph and rejoice in his success as the happiest woman in the world.

Now it was all over. It was noted that Marconi had never visited the Holman home in Indianapolis. Henry McClure, acting as spokesman for Josephine and her mother, said they had sailed for Europe, but their names could not be found on any passenger lists. Reporters cornered Marconi, who reluctantly gave a very brief interview. According to the *Daily News* he 'appeared disturbed and depressed', and was 'unwilling to discuss the motives for Miss Holman's action'. He confirmed that it was she who had called off the engagement in a letter to him, and that he had accepted in reply. Asked whether she had grown tired of waiting while he spent all his time trying to send a signal across the Atlantic, Marconi replied: 'The progress of my experiments has been greatly delayed by adverse fortune, and this had much to do with the delay of my marriage arrangements. But there was also a very delicate question involved, which is more than I have admitted to any other newspaper.' He refused to say what this 'delicate question' was. In an earlier interview with another newspaper Miss Holman said, 'There have been disasters on both sides,' but she too would not explain what she meant.

The *Daily News* speculated that Marconi's family were socially ambitious for him, and had not approved of Miss Holman. Perhaps neither family was happy about the way the two had met. 'It was a love match, pure and simple, born of the dangerous propinquity of all ocean steamship acquaintance,' opined the *News* reporter. Friends and relatives made the point that the two lovers had seen very little of each other as Marconi was 'wedded to his other mistress', science.

The loss of his fiancée clearly upset Marconi, but it did not interrupt his work, and when he left New York on 22 January 1902 on the American Line's SS *Philadelphia* he was already planning his next wireless coup.

18

Farewell the Pigeon Post

In the summer of 1902 the most eagerly awaited sporting event in America, and indeed throughout much of the world, was a boxing match to be staged in San Francisco on 25 July between two celebrated heavyweights, Bob Fitzsimmons and Jim Jeffries. Fitzsimmons was from the Cornish town of Helston, close to Marconi's station at Poldhu. He had arrived in San Francisco in 1890, and knocked out Jack Dempsey in New Orleans in 1891. So great was his celebrity that he had appeared in vaudeville and a Broadway play. In 1899 he had been knocked out by Jeffries, who is still considered by many to be the greatest fighter of all time. The San Francisco fight was the long-anticipated rematch. Would Fitzsimmons, nearing the end of a long career in which he was reputed to have stepped into the ring for 350 professional bouts, be able to outpunch Jeffries? The result of the fight would be cabled across continents and oceans instantly, and would be flashed to liners out in the Atlantic that had wireless equipment.

One or two places, however, would have to wait to hear the outcome, because they had no cable link. Among them was Santa Catalina Island, which lies twenty-two miles off the Californian coast. The islanders were as eager as anyone else to know the result, but they would not get it until the morning mailboat brought the newspapers. Their only other form of communication with the mainland was the intermittent pigeon post run by a hotel, and that was not flying on 25 July. The residents of the main town of Avalon, many of whom had gambled on the outcome of the fight, expected as always to wait for the news.

It arrived, however, much sooner than they had expected or could believe. Around midnight a notice appeared on a billboard in Avalon saying that Jeffries had knocked Fitzsimmons out in the eighth round. It had been sent from the mainland by wireless. There had been heavy betting on the fight, but nobody would pay up on the evidence of this advance news. Even when the newspapers confirmed the result, rumours abounded on the island that false claims had been made about the wireless report. Some said a man had been seen arriving in a boat at around the time the bulletin went up. Others said a carrier pigeon must have been used. The most popular story was that a strong semaphore light had been used: searchlights signalling sporting results were not uncommon. The *Los Angeles Times* had no doubt that trickery was involved, even though the distance covered by the wireless station was no more than that Marconi had achieved in 1898 with text messages from the royal yacht *Osborne* to the Isle of Wight. It took an investigation by a rival paper, the *Los Angeles Herald*, to establish that there was a real, working station in Avalon, and another on the Californian coast, providing the first ever regular wireless news service in America.

The little stations had been set up by a young man called Robert H. Marriott, who had become interested in wireless while studying physics at university just at the time, in 1897, when X-rays had been discovered and there was also great interest in Marconi's invention. His Professor had recently burned himself very badly while experimenting with X-rays – the dangers of radiation were not at first understood – and advised Marriott to try wireless instead. Over the next three years there was enough information in the technical magazines for Marriott to build himself transmitters and receivers, although the apparatus he had was very crude. He was so enthralled by the new technology that he took a job with the newly formed American Wireless Telephone and Telegraph

Company before he had finished his studies. In its early days the company was almost entirely fraudulent, intent only on extracting money from ignorant investors, and had no intention of setting up operating wireless systems. Its patents were those of Professor Dolbear, whose induction system was worthless. One of Marriott's first tasks for the company was to attempt to sabotage the coverage of the 1901 America's Cup, which Marconi's engineers and Lee de Forest were covering.

Although the American Wireless Telephone and Telegraph Company gave little evidence that it took its business seriously, one or two of its subsidiaries did. Among them was the Pacific concession which was established in Denver, Colorado, where Marriott was based. He and some other employees were aware that Dolbear's system was no good, and put together their own equipment which combined elements of the inventions of Reginald Fessenden and Marconi. Exactly how Marriott's detectors work is not clear, but they involved at one time a piece of old tin can 'oxidised' with a blowlamp, and all kinds of bits and pieces put together. It is likely that the detector was a copy of Fessenden's 'barreter', and was certainly not a coherer of the kind used by Marconi. The dots and dashes were received in earphones, which made the speed of sending and writing out messages painfully slow, as neither Marriott nor his colleagues were trained operators.

They decided that a link between Santa Catalina and Los Angeles would be practicable and worthwhile, and had a rudimentary system working by early July 1902. They chose the Fitzsimmons–Jeffries fight to make their first attempt to impress sceptical islanders, and the newspapers on the mainland, that they could send messages to Catalina. The station was soon to prove its worth. Two men who had robbed the island's Metropole Hotel of some champagne and other drinks took the 5 a.m. boat, hoping to make their escape before the theft was known about on the mainland. However, the Avalon station sent a wireless message to the police, who arrested them as they docked. This was the first instance in history of the use of wireless to apprehend criminals on the run. Little by little the potential uses of wireless were becoming apparent. But not everyone was yet convinced it was any more than some kind of magic show.

The scattered subsidiaries of the bogus American Wireless Telephone and Telegraph Company gave a number of intriguing demonstrations of the possible uses of wireless. A Kentucky branch, for example, acquired the patent rights in a device created by a farmer and telephone repairman, Nathan B. Stubblefield, and in return offered him a tranche of its company shares. For his first public demonstration, on 1 January 1902, Stubblefield and his son Bernard drew a crowd outside the Calloway County courthouse in Kentucky. Father and son set up two boxes on the ground, about two hundred feet apart. Young Bernard then took out his harmonica and began to play into one of the boxes, far enough away to be inaudible to the crowd. Father Nathan held to his ear a telephone receiver attached to the other box, and showed the audience that he could pick up Bernard's rendition of popular tunes. This performance must count as one of the earliest wireless broadcasts in history, but few at the time believed it was anything other than a conjuring trick.

However, the St Louis *Post Dispatch* was sufficiently intrigued by the story to send a reporter down to Stubblefield's farm outside the town of Murray to check it out. The reporter took one of Stubblefield's boxes and carried it about a mile from the farm. He was told to prod two iron rods into the ground when he set it up. To his astonishment he could hear the voices of Stubblefield and his son through the receiver, and heard Bernard playing the harmonica. With the endorsement of the *Post Dispatch*, Stubblefield was invited to demonstrate his wireless telephone in Washington, DC. Here one of his boxes was put on a steamer on the Potomac River, and others on the shore. Again they worked splendidly, and another demonstration was given in Philadelphia.

But the American Wireless Telephone and Telegraph Company never applied for the rights to the patents Stubblefield had agreed to assign it in return for its worthless shares. The company was

uninterested in advancing wireless technology – it was in business
to fleece the public with empty promises of future profits. Nothing
came of the farmer's invention. However a legend arose that it was
Stubblefield, an amateur electrician way ahead of his time, who
had invented radio in 1902. In time his farm became part of Murray
State College in Kentucky, and today there is a touching memorial
on the campus which reads:

Here in 1902
Nathan B. Stubblefield
1860–1928
inventor of radio-broadcast received the human voice
by wireless. He made experiments 10 years earlier
His home was 100 feet west

But Stubblefield's magic boxes were not radio: they worked on
the same principles as Preece's wireless telegraphy, that is, by induction rather than electro-magnetic waves. This is evident from
patents he applied for himself, and which he was granted in the
United States and Canada after leaving the American Wireless
Telephone and Telegraph Company. Stubblefield died embittered
and impoverished, and the company he felt had swindled him was
eventually convicted of fraud.

Stubblefield and his son had thought of their brand of 'wireless'
as a means of public broadcast, however down-home and amateur-
ish their public demonstrations. This had not occurred to Marconi
or any other leading inventors of the time. Stubblefield did have
his brief hour of glory, when the *Washington Times* reported that
'in basic form he has achieved something even greater than Marconi'. It was not at all clear at the time what wireless could be used
for, apart from its unique ability to set up communication to and
from ships or isolated islands which had no telegraph cable connection. When Marconi's Atlantic triumph was first written up in the
American press there were many cartoons which showed the cable
companies facing defeat, Goliaths challenged by the David of wireless. The United States was by then strewn with telephone and

telegraph wires, and it was anticipated with pleasure that these eyesores would soon disappear. There were bits of doggerel written about the dilemma birds would face when their telegraph perches were gone. Nearly everybody saw wireless simply as a competitor for the telegraph cable. It might possibly carry telephone calls, if the problem of sending speech through the ether could be solved. But nobody foresaw its development as a means of mass communication.

19

❧❧❧❧❦❦❦❦

The Power of Darkness

On his return journey across the Atlantic from New York in January 1902, Marconi took two of the fourteen luxury suites on the *Philadelphia*, which was not quite a state-of-the-art liner, but splendid enough. The cabins had electric light and hot running water, and Marconi had a brass bed, a sitting room furnished as in a private house, and his own bath and lavatory. Ingenious fans allowed air into the magnificent staterooms while keeping the salt water out. The ship's grand dining saloon was fifty-three feet long, and light filtered in through an arched glass roof twenty-five feet high. From their swivel chairs passengers could admire the murals of mermaids and dolphins. There was an oak-panelled library with stained-glass windows on which poems about the sea were written, and smoking rooms in black walnut, with scarlet leather sofas and chairs. For those genteel passengers who had not got their sea-legs, or who were too old or weary to climb the staircases, there were 'constantly ascending and descending electric chambers'.

Before the *Philadelphia* sailed, Marconi had arranged for engineers based at his newly formed American company to fit it out with wireless, so he could communicate with shore stations at Nantucket as they steamed away from the east coast of the United States, and pick up signals from Poldhu and the other stations as he approached the English south coast. The faithful George Kemp made sure Marconi ate regular meals, ordering breakfast for 8 a.m.

prompt, always two boiled eggs, toast and marmalade and tea. On the crossing to England Marconi got to know the captain and crew of the *Philadelphia*, and began to formulate a plan which, if agreed to by his board of directors, would silence his critics and demonstrate beyond any doubt that he was not fooling himself and the rest of the world with his claim to have heard the letter 'S' sent from Poldhu to St John's.

In the three weeks Marconi was back in England his company got agreement from the *Philadelphia*'s owners to heighten the ship's masts, and the tireless Kemp went to work. He noted in his diary: 'I fitted a four part aerial on three bamboo crosses made with 14 foot bamboos; one at the masthead, the second at the mizzen mast head and the third between the two nearest boat davits to the instrument room.' The receiver was set up in a cabin, and Marconi made a point of using one of his own coherers. It had been discovered that in Newfoundland he had used a different 'mercury' coherer developed by the Italian navy, and some critics had argued that this took away some of the personal glory of his achievement. He discarded too the telephone receiver he had used in St John's in favour of a Morse printer, so that there could be no doubt that the signals were being received, and were not a fluke or a figment of his imagination.

Captain A.R. Mills and the *Philadelphia*'s officers agreed to witness messages received as the ship crossed the Atlantic for New York, and to sign the Morse tapes to confirm where and when they had been received. Before the *Philadelphia* sailed from Southampton on 21 February, Marconi instructed the Poldhu station to transmit signals two hours out of every twelve, or one hour out of every six, in periods of ten minutes, alternating with intervals of five-minute rests. He would not be able to send messages back once the ship was more than 150 miles out, because his on-board transmitter did not have the necessary power. The liner called at Cherbourg on the northern French coast, and Kemp and Marconi spent the night there in a hotel. Kemp then returned to Southampton, wishing his boss good luck on the *Philadelphia*.

The crew were sceptical, as they had every right to be. It was of vital importance to Marconi that he should convince the captain and his senior officers that there was no trickery involved. For the uninitiated there was still something suspiciously magical about wireless. Rudyard Kipling had been inspired by Marconi's fame to write a short story which was published in *Scribner's* magazine in 1902. In the story, simply called 'Wireless', an amateur enthusiast is demonstrating how he can tune in to a station at Poole and pick up signals from Royal Navy ships at Portsmouth. A lot of the Morse messages get jumbled up, and the navy operators become irritable. The enthusiast says: 'That's one of them complaining now. Listen: "Disheartening – most disheartening." It's quite pathetic. Have you ever seen a spiritualistic seance? It reminds me of that sometimes – odds and ends of messages coming out of nowhere – a word here, a word here and there – no good at all.' Kipling was struck by the apparently supernatural quality of wireless, and in the story an elderly man falls into a daze and appears to pick up verses sent through the ether by the long-dead poet Keats, who has possibly been contacted by the 'Hertzian' waves of the amateur's little station.

Whatever the officers on the *Philadelphia* believed, they would have known that wireless had been shown to work over short distances, and was in use on other ships. But there was a persistent and widespread belief that its range had well-defined limits. And it would not have been difficult for Marconi to arrange for signals to be sent to his receiver from another ship on the Atlantic – or even from another part of the *Philadelphia* – and to pretend they had come from the Poldhu station. Such deceptions, with simple induction systems of wireless or hidden telephone links, had been practised by magicians and bogus spiritualist mediums for years.

On the second morning at sea, with the *Philadelphia* five hundred miles from land, an astonished Chief Officer Marsden saw and heard the tickertape of the Morse code printer rattle out a message which read: 'All in order. V.E.' The 'V.E.' was short for 'Do you understand?' Marsden told the captain what he had seen, but there was still scepticism. Marconi then asked the captain and officers to witness the receipt of messages from Poldhu at appointed times. The events which followed were described in a piece written by Henry McClure for *McClure's* magazine.

Watch in hand, Marconi sat looking at his instruments. Then he opened a brake on the coil of tape, and the white strip began to unroll. Suddenly Marconi burst out, 'There it comes,' and simultaneously began the 'tap, tap, tap' of the inker, and another message had waved itself through the ether, and had been recorded on a piece of narrow paper nearly 1,000 miles away from its source. The days following were full of suppressed excitement. Shortly after midnight on the 24th, amidst scores of signals, came the message:

'Fine here.'

The distance was then 1,032.3 miles. Another about the same time read:

'Thanks for telegram. Hope all are still well. Good luck.'

The supreme test for messages came when the ship was 1,551.5 miles from Poldhu.

At that time, just before the break of day, on the 25th came the message:

'All in order. Do you understand?'

'Let me show you just how accurately these instruments operate,' said Marconi to Captain Mills. (The ship was then in mid-ocean.) 'I will release the brake on the coil of tape just a few seconds before the appointed time, and we shall see when the signals begin, and whether they come as they should.'

Mr. Marconi and the captain held their watches. Ten seconds before the expected working period a snap of the brake set the coil in motion. The two waited with breathless expectancy. The captain had been one of the skeptics at the start. Now he was all confidence and enthusiasm.

Almost exactly on the second for which they were waiting there was a slight buzz near the coherer. Marconi lifted his hand. 'Tap, tap, tap,' sounded the inker as it clicked against the tape. The young man smiled. 'Is that proof enough, Captain?' For exactly ten minutes the signals continued in unbroken order.

As the *Philadelphia* steamed towards New York, Captain Mills signed the Morse tapes to confirm the exact position at sea where they had been received. The distance from Poldhu was marked on a map, and a full chart of the voyage was produced and signed by the captain and Chief Officer Marsden. From time to time Marconi would turn his receiver on when Poldhu was off air to show the captain that what they were receiving was not the effect of atmospheric electricity, or a signal from another ship. His receiver was tuned only to Poldhu – another liner crossing the Atlantic at the same time, the *Umbria*, equipped with a Marconi receiver tuned to a different wavelength, received no Poldhu signals. Captain Mills's signature appears on the Morse tape to confirm that the *Philadelphia* received short messages up to 1551.5 miles from Poldhu. Beyond that, Marconi was able to pick up only the simple 'S' signal. This time it was recorded on tape: 'Received on S.S. "Philadelphia," Lat. 42.1 N., Long. 47.23 W., distance 2,099 (two thousand and ninety-nine) statute miles from Poldhu. Capt. A.R. Mills.'

When Marconi was greeted by newspaper reporters in New York he held up two miles of Morse printer tape. He was unusually unguarded and ebullient. It would not be long, he declared, before he had a transatlantic wireless system working; and he believed he would be able to send a signal right around the world, and pick it up at the point from which it had been sent. He told Henry McClure:

I knew the signals would come up to 2,100 miles, because I had fitted the instruments to work to that distance . . . if they had not come, I should have known that my operators at Poldhu were not doing their duty.

Why, I can sit down now and figure out just how much power, and what equipment would be required to send messages from Cornwall to the Cape of Good Hope or to Australia. I cannot understand why the scientists do not see this thing as I do. It is perfectly simple, and depends merely on the height of wire used and the amount of power at the transmitting ends. Supposing you wanted to light a circuit of 1,000 electric lamps. You would use enough dynamos and produce enough current for that effect. If you did not have that much power, you could not operate 1,000 lamps. It is the same with my system. We found several years ago that, if we doubled the height of our aërial wire, we quadrupled the effect. We used one-fortieth of a horse-power then. Now I use several horsepower, and, by producing a powerful voltage, I naturally get an effect in proportion to that power. It is not possible to keep on extending the height of our aërial conductors, so we simply use more power when we wish to do long distance work.

But there was one aspect of the *Philadelphia* tests which troubled Marconi, and which he either did not mention to McClure, or the reporter chose to leave out of his story. The distance at which signals could be received from Poldhu was much greater at night than during the day. When the liner was seven hundred miles out from Southampton, all daytime signals were lost. Marconi thought it might have something to do with the effect of sunlight on the aerials he was using, but he knew of no solution to the problem. Cable was unaffected by light or darkness, but wireless clearly was, and this could be a fatal blow for a rival transatlantic service.

Marconi's dilemma was in fact the clue that science needed to solve the riddle of his magic box. But no one could explain it. Oliver Lodge, Nikola Tesla, Sir Ambrose Fleming – in fact everyone working with wireless – still believed that electro-magnetic waves travelled through a substance as mysterious as Lodge's departed spirits: the ether. But how it was that they could travel so far was no better understood than it had been when Marconi first showed that they could be detected a mile away. It would take a mind of real but troubled genius to provide the answer.

departed from the ether. But how it was that they could travel so far was no better understood than it had been when Marconi first showed that they could be detected a mile away. It would take a mind gifted but troubled genius to provide the answer.

20

❧❧❧❧❧

The Hermit of Paignton

Roughly halfway between the Haven Hotel in Dorset and Poldhu in Cornwall, on that stretch of Devon's south coast which likes to call itself 'the English Riviera', is the resort of Torquay. In the spring of 1902 an eccentric Englishman would sometimes be seen flying down the steep country lanes outside the town on a bicycle, followed by an anxious, and not quite so daring, companion. They called this reckless cycling 'scorching', and the more timid man, George Searle, a Professor of Physics at Cambridge University, recalled: 'We used to put our feet on the foot-rests on the front forks and then let the cycle run down hill. Oliver put his feet up, folded his arms, and let the thing rip down steep and quite rough lanes, leaving me far behind.' This was one of the few pleasures in the mostly unhappy life of Oliver Heaviside, who was known locally and unflatteringly as 'the Hermit of Paignton', the small town nearby in which he had lived for a time.

Heaviside was born in 1850 in Camden Town, then a rough area of London close to where Charles Dickens had spent the unhappiest part of his childhood. The family was poverty-stricken, for Oliver's father was a wood engraver and the development of photography from the 1850s had ruined his trade. He was a violent man, and regularly beat Oliver and his brothers. The Heavisides were saved from their penury when a sister of Oliver's mother married Charles Wheatstone, a wealthy man who had been closely involved with the development of the electric telegraph. Oliver was largely self-educated, having left school at the age of sixteen to work in a telegraphy office. He spent some time in Denmark working for a cable company based in Newcastle-upon-Tyne, but gave up work entirely at the age of twenty-four to study the mathematics of James Clerk Maxwell, whose theory of electro-magnetism had led to Marconi's creation of a practical wireless system.

Charles Wheatstone gave Oliver an interest in music as well as science, and through his influence two of Oliver's older brothers obtained jobs with a telegraph company. The family were still in Camden Town when Oliver gave up his job in favour of sitting alone in a small attic room, smoking a pipe and working his way through Maxwell's mathematical equations. He later moved to Torquay to stay with his brother Charles, who managed a music shop there. Oliver Heaviside began his work in 1874, the year Marconi was born, and two years later published in the *Philosophical Magazine* a highly technical paper concerned with a recurrent problem in cable telegraphy, about which he knew a great deal. It was quite unintelligible to the lay reader, but those who could make sense of it recognised that Heaviside was quite brilliant. For the rest of his life he conducted a debate with the leading physicists and electrical experts of the day, either by writing to them or by publishing articles in the magazine *The Electrician*. These were often vitriolic, and many were aimed at the man he disparaged most: William Preece of the Post Office, Guglielmo Marconi's erstwhile benefactor. Heaviside regarded Preece as a mathematical and scientific incompetent who talked and wrote nonsense, and understood nothing about telegraphy.

Heaviside's life was utterly different from Marconi's. In an affectionate memoir Professor G.F.C. Searle recalled a visit he made with his wife to Heaviside's home: 'We had been warned what we should find. The teapot spout was completely stopped up by tea leaves and no tea could come out of it. Oliver tipped the pot so far that tea ran out of the top. He caught what he could in the cups, and carefully spooned the tea leaves out of Mrs Searle's cup.'

On another visit Heaviside told them: 'There are nine pieces of bread and butter – three pieces each. There is some cake at the end but I would not recommend it.' Heaviside wrote to Searle around 1902: 'if I boil an egg, I am startled by a loud report; either I did not put any water in or else it has boiled away'.

At that time Heaviside was fifty-two years old, living in squalor with a woman relative who acted as housekeeper. His strange manner drew the attention of local boys, who constantly persecuted him, throwing stones at his windows. He offered the local police 150 apples from his garden if they would keep a watch on his home. And yet this social misfit was one of only two scientists in the world in 1902 who proposed the correct solution to the puzzle of how Marconi was able to transmit wireless signals to distances of over two thousand miles.

Heaviside was a pure scientist: he arrived at Heinrich Hertz's proof of electro-magnetic waves through deduction at the time the German was testing James Clerk Maxwell's ideas in his laboratory. When Marconi's wireless telegraphy began to receive wide publicity, Heaviside was quick to point out that William Preece – who was publicly suggesting it was he, and not Marconi, who had developed it – did not understand it at all, and could not tell the difference between induction and Hertzian waves. In September 1900, while Marconi and Ambrose Fleming were about to start work at Poldhu, Preece was telling the British Association: 'The sensation created in 1897 by Mr Marconi's application of Hertzian waves distracted attention from the more practical, simpler and older method.' By 1907 he was telling a committee of the House of Commons that he had been working on wireless telegraphy twelve years before Marconi came to Britain.

When Heaviside heard about Marconi's transatlantic signal he quickly had an explanation based on his profound understanding of electro-magnetism. He wrote to *The Electrician* suggesting that there might be a conducting layer in the upper atmosphere which reflected the wireless waves, but his letter was not published. Then, in 1902, he was asked to contribute a piece on 'The Theory of the Electric Telegraph' to the tenth edition of the *Encyclopaedia Britannica*. In it he mentioned, almost in passing, that seawater had 'enough conductivity' for Hertzian waves to bounce off it, and that 'there may possibly be a sufficiently conducting layer in the upper air. If so, the waves will, so to speak, catch on it more or less. Then the guidance will be by the sea on one side and the upper layer on the other.'

The theory of an upper layer of the atmosphere which reflected wireless waves downwards was also proposed by Arthur Kennelly, an expatriate Briton working in America. But nobody took any notice, and there were plenty of rival theories. None of this mattered much to Marconi, for even if there *was* a reflecting layer sending his long wireless waves back to earth, nobody had any idea how it behaved.

Oliver Heaviside turned down a number of honours. He declined the invitation to the banquet held for all the contributors to the *Encyclopaedia Britannica*, although he did accept the fee. In his last years he took to awarding himself the self-deprecating acronym 'W.O.R.M.' after his name. There is no record of any correspondence or contact between him and Marconi.

As far as the development of wireless went, Heaviside's brilliant speculation was of no practical use at the time, because nobody knew how to examine the properties of the upper atmosphere. And there were still many urgent unanswered questions. Would wireless waves travel as far over land as over the sea? Most self-appointed experts thought not; mountain ranges might prove to be an obstacle. Marconi himself was not sure. There was only one way to find out.

21

The King's Appendix

Queen Victoria had died at Osborne House on the Isle of Wight on 22 January 1901. The nation was in mourning for six months, decked out in miles of black crêpe. The coronation of her son Edward in Westminster Abbey was fixed for 26 June 1902, and for weeks beforehand London was transformed, with the arrival of exotic troops of soldiers and sailors from all over the Empire who would take part in the coronation parades. There was to be a naval review at Spithead, and King Victor Emmanuel III of Italy made arrangements to arrive by sea aboard his cruiser the *Carlo Alberto*. The twenty-nine-year-old Emmanuel had come to the throne only two years earlier, after the assassination of his father, King Umberto, by a member of an international group of anarchists, an Italian from New Jersey.

A friend of Marconi's, the Italian naval officer Luigi Solari, had asked the young King if a ship might be made available for wireless experiments. Fiercely proud that the world-famous inventor of wireless was an Italian national, Emmanuel invited Marconi to install his equipment aboard the *Carlo Alberto*, so that when the coronation and naval review were over Marconi could join the King and conduct whatever wireless tests he liked as they steamed back to Italy. This was a tremendous opportunity for Marconi and he happily accepted Emmanuel's offer.

When the *Carlo Alberto* came in sight of the English coast on the morning of 18 June 1902, the captain, Admiral Mirabello, the King and the ship's crew could see the masts of Poldhu in the distance. With equipment Luigi Solari had fitted to the ship they exchanged messages with the Marconi station. The next day George Kemp bought a large Union Jack in preparation for Coronation Day, and hoisted it on one of the wireless masts.

At 11.15 a.m. on 24 June a notice was posted outside Buckingham Palace which read: 'The King is suffering from perityphlitis. His condition on Saturday was so satisfactory that it was hoped that with care his Majesty would be able to go through the Coronation ceremony. On Monday evening a recrudescence became manifest, rendering a surgical operation necessary today.' Crowds gathered outside the palace waiting for news, while others went to St James's Square to see if they could learn anything from the Earl Marshal at Norfolk House, organisational headquarters for the coronation. *The Times* reported: 'But the servant who answered the door . . . had nothing to say to the general inquirer. Even to the accredited press representative his answer was courteous but laconic: "The Coronation will be postponed; we can tell you no more." '

In early-twentieth-century England, only the poor were treated in hospitals: the well-to-do were attended and operated on at home. Edward had what we now call appendicitis. He needed an operation, and the surgeon chosen was the celebrated Frederick Treves of the London Hospital, who had announced his retirement in his early forties after making a fortune in private practice. It was Treves who had befriended and found a haven for the deformed Joseph Merrick, known as 'the Elephant Man'. An operating bed was brought from the London Hospital and installed in a room in Buckingham Palace, with one nurse in attendance.

The anaesthetics in use at the time were ether and chloroform, which were applied only briefly. Surgeons like Treves prided themselves on the speed with which they worked so patients were not rendered unconscious for long. Edward, however, was sixty-four years old, overweight and not in the best physical condition. He reacted violently to the first administration of the anaesthetic, and almost swallowed his tongue: his beard had to be pulled hard to retrieve it so that he did not choke. Queen Alexandra, who was present, became hysterical and had to be ushered from the room. Treves, wearing a dirty old coat – he was an old-fashioned practitioner, with little time for what he regarded as the 'continental' obsession with cleanliness which had developed after the discovery of micro-organisms – then proceeded to open up the future monarch and drain the pus from his infected appendix.

It was not at all certain that Edward would survive, and if he did, how long the coronation would be delayed. Many foreign visitors packed up and went home. King Victor Emmanuel decided to go back to Italy on the *Carlo Alberto*, and hasty arrangements were made to get Marconi's equipment aboard at Poole. Then Emmanuel changed his mind. He had been in touch by cable with Tsar Nicholas, who invited him to Russia. Marconi had gone to London, leaving George Kemp to organise the fitting-out of the *Carlo Alberto*. At the last minute it was arranged that he would meet the ship at Dover, by which time Kemp would have it ready for an exciting series of experiments as it sailed north. An extra mast was raised on the ship, and a four-stranded aerial rigged as high as possible as the *Carlo Alberto* steamed towards Dover. Once again Kemp stayed behind in England, waving Marconi goodbye and wishing him well with his latest invention. On board the *Carlo Alberto* for this unexpected voyage north was an entirely new kind of detector which Marconi had only just put together.

Any device which reacted to the impact of electro-magnetic waves could be turned into a receiver of Morse dots and dashes. Marconi had read accounts by the brilliant New Zealand physicist Ernest Rutherford of experiments in which electro-magnetic waves were used to demagnetise iron needles. Just three years older than Marconi, Rutherford was a true scientist who had become famous for his work on radioactivity; in 1931 he would be honoured with the title Lord Rutherford.

In 1894 Rutherford had written a paper showing how a coil of magnetised wire could become a detector of wireless waves. It took Marconi's dogged determination and craftsman's ingenuity to fashion from Rutherford's blueprint a practical wireless detector out of bits and pieces he could lay his hands on while working at the Haven Hotel. Like the Vail brothers, who bought milliners' copper wire designed for 'skyscraper hats' for their experimental electric telegraph, Marconi made use of an unexpected source for the thin wire he needed. While cycling around the Poole area he had often gone into Bournemouth, and his attention had been attracted by a pretty girl in a flower shop who worked all day making posies for visitors to the resort. Marconi remembered that she had used a delicate wire frame for their structure, and in a moment of inspiration he cycled to Bournemouth and persuaded the girl in the flower shop to sell him some lengths of fine decorative wire. He almost 'scorched' Heaviside-fashion back to the Haven Hotel in his impatience to begin winding it into a coil.

George Kemp produced a wooden Havana cigar box in which to house the new receiver, and horseshoe magnets were not hard to find. But Marconi also required a revolving strand of wire on two small wooden spools, and these needed a miniature motor of some kind. As always, late-Victorian ingenuity unwittingly provided just the thing. On 2 June Kemp noted in his diary: 'I went to Bournemouth where I bought a second-hand Edison Phonograph; the clockwork was taken out and used for revolving the iron core of No. 40 silk covered iron wire through the primary coil of the detector.' Cannibalised Edison cylindrical phonographs were to provide several of the early motors for Marconi's new invention.

As the *Carlo Alberto* raced up the North Sea, Marconi worked night and day testing his magnetic detector. It worked well apart from a persistent problem with atmospheric interference, the price of its greater sensitivity. He also rigged up a coherer and Morse printer to provide a permanent record of messages received. But once again, as on the *Philadelphia*, the sunrise cut them off from Poldhu once they were five hundred miles away, and signals could not get through again until half an hour after sunset. With a midshipman, Raineri-Biscia, Marconi worked through the night adjusting his equipment and trying out all kinds of devices which he hoped might bring Poldhu back in daylight. But nothing worked. Raineri-Biscia noticed that Marconi became more and more agitated, shouting in Italian with a Bolognese accent: 'Damn the sun! How long will it torment us!'

When the *Carlo Alberto* reached Russia they anchored at Kronstadt, and King Victor Emmanuel went off to meet Tsar Nicholas. Naturally the Tsar wanted to see for himself this remarkable invention of Marconi's, with which he was able to receive messages from Cornwall, 1600 miles away. He arrived in splendour, with the battleships of his fleet firing salutes, the sailors cheering and bands playing the national anthem. Marconi showed the wireless equipment to the Tsar, who was thrilled to see a special greeting tapped out on the printer. Speaking in English, he asked Marconi where it had come from. Marconi apologetically explained that it had been sent not from Cornwall, but from the other end of the *Carlo Alberto*, where a transmitter had been hastily put together by Solari. They would not hear from Poldhu again until the sun had set.

As a matter of courtesy, King Victor Emmanuel had arranged to rendezvous at sea with the German Emperor. Kaiser Wilhelm II, cousin to Edward VII, was regarded in royal and diplomatic circles as 'not quite sane'. Bismarck said of him: 'The Kaiser is like a balloon. If you do not hold fast to the string, you never know where he will be off to.' It was Wilhelm who had demanded that Germany have great ships like the British, and who had ordered the building of the *Kaiser Wilhelm der Grosse* and the *Deutschland*, both former holders of the Blue Riband. It was also he who had asked William Preece to allow Professor Adolphus Slaby to spy on Marconi. The Kaiser wanted Germany to be ahead of all other countries in new technologies, and it did not please him that Marconi appeared to be far more successful than his own scientists and inventors. The wireless system which Slaby and Count von Arco had developed was fitted on a few ships by 1902, one of them the *Deutschland*. Other German vessels, like the *Kronprinz Wilhelm*, had Marconi equipment. And it was the Marconi Company that had most of the shore stations.

Early in 1902 the Kaiser's brother, Prince Heinrich, had made an official visit to the United States, sailing on the *Kronprinz Wilhelm*. He was pleased to find that he could send messages via Marconi operators either to the east or the west all voyage, and was never out of wireless contact. He sailed back from America on the *Deutschland*, and found to his great annoyance that he could not get any messages through at all with its German wireless equipment. The Germans believed that the Marconi Company was refusing to listen in to Slaby-Arco operators because it wanted a worldwide monopoly of wireless. When the Kaiser learned of this apparent snub to his brother, he was furious. What the American magazine *Electrical World* described as 'malignant Marconiphobia' spread across Germany, and soon Slaby and others were writing indignant letters to the *New York Herald* complaining that they were the victims of deliberate wireless sabotage.

Marconi replied in letters to various newspapers, including the *New York Times*, that the *Deutschland*'s Slaby-Arco equipment was not tuned to his. It was simply a technical matter, and had nothing to do with any monopolistic ambitions of his company. Had there been the desire, the two systems could have been made compatible, though Marconi's was superior. His claim that the problem was technical rather than political was disingenuous: he was just trying to calm things down. Nevertheless, according to *Electrical World* the battle raged with 'berserk fury' in Germany, as it did in the columns of American newspapers. This grievance was still raw when King Victor Emmanuel ordered Admiral Mirabello to take the *Carlo Alberto* to the German port of Kiel for an official meeting with the Kaiser, who was wandering aimlessly at sea in the imperial yacht *Hohenzollern* awaiting the outcome of Edward's operation at Buckingham Palace.

Poor George Kemp, who had been working on installations in England, was cabled by Marconi with a message to get to Kiel as quickly as possible. At 9.25 a.m. on 22 July he bought a train-boat ticket to Kiel at Holborn station in London, and he was on a steamer heading for Germany at 11.30 that morning. Some of his luggage had gone missing when he reached the German coast and he had to cable for it before settling down in a restaurant car on his way to Hamburg. He was in Kiel soon after 11 o'clock the following morning – just over twenty-four hours after leaving London. As there was no sign of the *Carlo Alberto* he booked himself into a hotel, and watched the harbour. The next morning he noted in his diary: 'I was aroused by a band and a company of soldiers on the march.' German militarism was even then evident, a portent of the great conflict to come. Kemp watched the German battleships leave the harbour, and in the early afternoon was pleased to see the *Carlo Alberto* arrive. After a stroll along the seafront with Marconi and Admiral Mirabello he was given a cabin and dinner on the Italian ship.

The rendezvous with the Kaiser was some days off, and the time was spent testing Marconi's equipment in Kiel harbour. Kemp and Marconi stayed up until the early hours of the morning, as that was the only time they could pick up the Poldhu signals – half an hour after daybreak the signals would cease, which Kemp believed had something to do with a 'change in the earth's magnetic medium'.

The arrival of the *Hohenzollern* with an already aggrieved Kaiser aboard did not go well. It was around midnight, and Admiral Mirabello was ordered by the Germans to greet their King with a twenty-one-gun salute. He replied that it would be a breach of his orders to fire a salute after dark, and when the demand was repeated the decision was taken for the *Carlo Alberto* to leave immediately. According to the story told by Luigi Solari, which perhaps has some romance in it, as they passed the *Hohenzollern* they sent her a message by wireless, asking if she wanted news from Poldhu. The reply came back that such a long-distance signal was impossible, whereupon the German operator leaned on his key, jamming the airwaves with meaningless dots and dashes.

If such an incident did take place, it was an early salvo in what was to be a series of battles between the Marconi Company and the German Kaiser up to and during the First World War. Wilhelm ordered all German ships, including the navy, to use only Slaby-Arco equipment, even though it had a very limited range and was obviously inferior to Marconi's. He then began an international campaign to try to compel the Marconi Company to communicate with other wireless systems. A series of conferences were called to thrash out the issue, and international agreements were signed seeking to force Marconi to comply with the principle of intercommunication. These were resisted for several years, but the more Marconi demonstrated to the wonder of the world that wireless waves could carry Morse messages over huge distances, and the more reliable his system became, the greater the threat he faced from competition and government control.

Edward VII survived surgery at the hands of Frederick Treves, and after a period of convalescence his coronation was rescheduled for 6 August. The *Carlo Alberto* returned to England and had pride of place at the Spithead naval review, only to lose it to an unfortunate change in the wind which dislodged its anchor so that it was left trailing the field. After the celebrations King Victor Emmanuel visited Poldhu for a few days, and some of his officers went ashore to take a look at the amazing contrivances of Professor Fleming, and the giant web of a mast. Finally they set sail for Italy, with Marconi and Solari testing coherers and the new magnetic detector all the way. The crucial time for Marconi would be when the *Carlo Alberto* steamed into the Straits of Gibraltar, where the great landmass of the Iberian Peninsula would form a barrier between her and Poldhu.

Admiral Mirabello slowed the *Carlo Alberto* down in anticipation of the first attempt to pick up Cornwall at 2 o'clock on the morning of 5 September. The ship drifted in thick fog while Marconi, Solari and the Admiral waited by the receiver with earphones and Morse printer ready. Nothing came through. When the prearranged time for transmissions was up, Marconi paced the deck. There was no way he could find out if anything had gone wrong at Poldhu, and he had to wait until 3 a.m. before attempting to tune in again.

The *Carlo Alberto* continued to circle in the fog so that Marconi could have another chance to pick up Poldhu. When, at just after 3 a.m., the Morse printer began to tap out a series of 'V's – the standard call signal from Cornwall – Solari ran straight to the bridge to tell Mirabello. He found the cruiser lit up by a beam of light from another ship which had come out from Gibraltar to investigate. The *Carlo Alberto* had stayed around too long, and the British Navy considered its strange behaviour suspicious. Rather than involve himself in a dispute, Mirabello headed full steam for Italy.

Although he was once again triumphant, Marconi fell ill. He had had a year of incredible nervous anticipation and endless travel, and now he lay in bed with a high fever. When the cruiser was nearing La Spezia he rallied, and began to set up his equipment to receive a telegram he had arranged to have sent from Cornwall to King Victor Emmanuel aboard the *Carlo Alberto*. The signal came through, but when Marconi tried to decipher it, he found it was gibberish. In frustration and fury, he smashed up his receivers, believing that the operator of the Morse key at Poldhu must be incompetent. When he had calmed down his equipment was reassembled, but no signal came through. According to Solari, Marconi guessed that Poldhu had inadvertently changed wavelengths. The measurement of wavebands was still crude, and a slight alteration in the transmitter set-up required an adjustment to the receiver which was a matter of guesswork. To adapt his receiver Marconi wound some wire around a candle, and attached it to the aerial. Finally the message got through, but wavelength and tuning were still tormenting Marconi. And time, as always, seemed to be against him.

On the surface, however, everything was splendid. Victor Emmanuel had honoured Marconi with an audience, during which they talked over the excitements of the *Carlo Alberto* voyage. Marconi visited Bologna, where crowds cheered him, and he was given the use of an ocean-going steamer and her crew.

By the middle of September the *Carlo Alberto* was back in the English Channel, with Marconi aboard. He had been lent it by the Italian navy, as the King believed that its presence in Canada when the first regular transatlantic service began would give Italy a great deal of prestige. After a short stay in England Marconi headed back across the Atlantic to Glace Bay, Nova Scotia, confident that he would soon be sending the first readable Morse messages across the Atlantic both from east to west and west to east.

22

❧❧❧❦❦❦

The Thundering Professor

The transmitter at Poldhu built by Professor Fleming presented a frightening spectacle. Arthur Blok, who worked with Fleming in the early days, recalled:

> The eerie and alarming appearance of that spark in the rural background of Mullion and Helston is something not to be forgotten. When the door of the enclosure was opened, the roar of the discharge could be heard for miles along the coast. The local ether storm produced by this smashing discharge was also noteworthy. Every metal gutter, drainpipe or other object about the sheds on the site resonated freely and there was a minor chorus of ticks and flashes in consonance with the discharge. A sizeable spark could be drawn to the knuckle presented to a bunch of keys placed on the ground outside the discharger hut and when one climbed up a short wooden ladder that leant against the hut there was a tingling sensation whenever one's hand passed over the nails which secured the rungs.

The basic technology of generating wireless signals had not changed since Marconi first arrived in London with his magic boxes. To achieve greater distances it had merely grown, requiring massive power which could produce terrifying discharges of electricity. In contrast to Marconi's neat little cigar-box magnetic detectors, the transmitters were fire-breathing monsters. The sparks that were intended to send messages across the Atlantic were created between two revolving metal discs the size of dinner plates, and the power was stored in banks of giant Leyden jar batteries which together weighed tens of tons. Ambrose Fleming was then in his early fifties, and a little deaf.* But even he could hear Poldhu when it was transmitting, and anyone in the Poldhu Hotel who could translate Morse code by ear would have known what messages were being sent by the thunderous alternating dots and dashes.

Fleming had chosen the call-sign 'V' when testing the equipment, which in Morse code is 'dot-dot-dot-dash'. Blok recalled the Professor humming 'da-da-da-daaaa' to himself, or whistling it between his teeth. He had made a little machine which would fire off the letter 'V' continuously from a revolving tape, driving those around him mad. Fleming was a tireless researcher, constantly trying to improve Marconi's equipment with experiments both in Cornwall and at University College, London. In October 1902 he was preparing for the test which would show whether or not a transatlantic wireless service was possible. He had found a way of producing more power, and everyone at Poldhu now knew that the only hope of repeating the success of the 'S' the previous December was to confine transmission to the hours of darkness. How the 'S' had ever got to St John's in daylight they never understood, and it was a long time before they were able to repeat the achievement.

On the last day of October the *Carlo Alberto* with Marconi aboard arrived at Glace Bay to a tremendous reception. A flotilla of boats came out to greet him, and reporters gathered around Marconi to ask when the first transatlantic messages would be sent.

---

* It is striking how many of those who were prominent in the field of early electronic sound were afflicted with poor hearing. Thomas Edison was partially deaf from boyhood, as was Oliver Heaviside, who had suffered from a bout of scarlet fever as a child. Alexander Graham Bell became interested in electronics through his work with the deaf and dumb, and a fascination with the changing pitch of speech. His wife had been totally deaf since the age of five, when she too had had scarlet fever.

The Glace Bay station, on a promontory called Table Head, was to be even more powerful than Poldhu, and Richard Vyvyan had been given a free hand in its design. Vyvyan was Marconi's most trusted engineer, though he was not always appreciated by the directors of the company, who complained about his sometimes boorish behaviour: he insisted on smoking his pipe in areas designated as no smoking. Vyvyan had been able to call upon an experienced labour force to put up the buildings and huge aerials of the Glace Bay station, for in the previous ten years Cape Breton Island had undergone a dramatic industrialisation. In 1900 a steelworks had been opened outside the principal town of Sydney, and to supply it with fuel coalmines had been dug on the Cape. Towns grew up around the mines, which also supplied coal to the rapidly growing industries of the eastern seaboard and on the Great Lakes. The massive Dominion Steel and Coal Company, set up by an international conglomerate, advertised for labour not only locally but in Europe, and the migrants flooded in. The people of Cape Breton abandoned their farms and fishing and made for the squalid, hastily built townships, where they could earn much higher wages than in their traditional industries. In just one decade, from 1891 to 1901, the population of Glace Bay had risen from 2459 to nearly seven thousand, and there were five collieries, each employing two thousand men, many of them drawn in from other regions of Cape Breton. Vyvyan's cosmopolitan workforce spoke more than half a dozen languages. At any time he would have between one and two hundred men labouring under him, including Italians, Poles, Native North Americans, and many from south-eastern Europe.

The station's four towers, made of pine and embedded in huge concrete blocks, were more than two hundred feet high. They supported cables from which the aerials were suspended. In the middle of the square formed by the four giant towers were the powerhouse and the operating house. The electrical generator was steam-driven, and fuelled by the abundant Cape Breton coal. While some of the labourers found places to live in the newly built streets of the mining towns, others had to make do with rude huts they constructed themselves until Cape Breton, with revenue from the taxes on coal, managed to catch up with the great influx of people and introduce some town planning.

The Glace Bay station was reached by a muddy road which led to a high barbed-wire fence. For Marconi and his staff a single-storey wooden house was built, with nine bedrooms, a parlour, dining room and sitting room which had a piano, which Marconi sometimes played in rare moments of relaxation. The people of Cape Breton, according to the newspapers, were very proud that such a famous figure had camped in their community, and showed great interest in the station, an interest that was often fuelled by rumour. Journalists were forever arriving in the hope of getting an interview with Richard Vyvyan, who rebuffed them with the warning that he was prepared to talk about anything but the workings of the heavily secured station and its equipment. On his days off he fished in the brook that ran through the station site, catching bags full of sea trout as they swam up the tidal estuary to their breeding grounds.

All appeared to have gone well with the construction of the station, but when Marconi told reporters he could give them no promises about when his much-vaunted transatlantic service would begin to operate, they were disappointed. All along, Marconi had feared that the expectations of his Canadian sponsors were wildly optimistic. Winter was setting in, draping the station in thick fog and obscuring the *Carlo Alberto*, which was moored in the bay awaiting news. And when Marconi began to get to grips with the equipment, there was only silence in his headphones. Fleming and his men were banging out thunderous streams of 'V's, but none reached Glace Bay. Fleming was cabled to step up the power, and the monstrous Morse signals echoed around Mullion Cove for twenty-nine days, without any signal getting through to Glace Bay. To add to everyone's anxiety, Marconi received a telegram from the board of directors telling him the company's share price was falling. Added to all this, at Christmas the *Carlo Alberto* would have to weigh anchor and head for South America.

On 19 November Glace Bay began to crash out signals which lit up the icy landscape, but nothing was received in Cornwall. For nine days, as the weather hardened, Marconi and others pressed the Morse keys without a result. On the night of 28 November Poldhu picked up signals, but they were weak and unreadable. Snow began to fall on the Glace Bay aerial masts, and soon lay thick on the ground around the transmitting station. Nothing was heard again on either side of the Atlantic until 5 December, when Glace Bay received news from Poldhu that they had received signals, most of which were weak, though some had been strong enough to activate the Morse printer. Another ten days of blindly adjusting equipment went by before the heart-stopping sound of clear signals was heard from Cornwall, telling them that for two hours they had had readable messages.

It was seven o'clock in the morning of 15 December 1902. The men who were working ran out into the snow in mad rejoicing. But not Marconi: he had to make a fateful decision. Once the board of directors knew of this small success they would want to announce that a commercial wireless telegraphy service between Cape Cod, Glace Bay and Poldhu was ready to go into action. It could not begin straight away, because the British Post Office still retained a monopoly on telegraphic messaging, and that problem would have to be sorted out. Left to make his own judgement, Marconi would have kept quiet and said there was more work to do. Richard Vyvyan would have backed him. Communication between the two stations was agonisingly slow and unpredictable. Some messages had to be sent and resent, up to twenty-four times in one case, before they could be deciphered at Poldhu.

But Marconi was now a servant of his own company. His magic boxes had carried him to a point of no return. Without any elation, he carried out a prearranged plan to send greetings from Glace Bay to Victor Emmanuel III in Italy and Edward VII in England. But first he would call in a reliable journalist to witness the sending and receiving of messages. There was still great uncertainty in the minds of many people about what Marconi had achieved in

Newfoundland and aboard the *Philadelphia*. It would have been so easy to pretend that those invisible messages had travelled more than two thousand miles when they had in fact been deceitfully transmitted from some nearby hideaway.

One newspaper of authority which had believed in Marconi in December 1901 was the London *Times*. In recognition of the faith it had shown then, Marconi chose George Parkin, the *Times* correspondent in Ottawa, to witness the messages sent to Poldhu which would be relayed by cable to London. Parkin later described the scene on the night of this historic transmission:

A little after midnight the whole party sat down to a light supper. Behind the cheerful table talk of the young men on the staff, one could feel the tension of an unusual anxiety as the moment approached for which they had worked, and to which they looked forward so long. It was about ten minutes to one when we left the cottage to proceed to the operating room. I believe I was the first outsider to inspect the building and the machinery.

It was a beautiful night – the moon shone brightly on the snow-covered ground. A wind, which all day had driven heavy breakers on the shore, had died away. The air was cold and clear. All the conditions seemed favourable. Inside the building and among its somewhat complicated appliances, the untechnical observer's first impression was that he was among men who understood their work. The machinery was carefully inspected, some adjustments made, and various orders carried out with trained alertness. Everyone put cotton wool in their ears to lessen the force of the electric concussion, which was not unlike the successive explosions of a Maxim gun. As the current was one of the most dangerous strength, those not engaged in the operations were assigned to places free of risk.

It had been agreed at the last moment before transmission I should make some verbal change in the message agreed on for the purpose of identification. This was now done and the message thus changed was handed to the inventor who placed it on a table where his eye could follow it readily. A brief order for the lights over the battery to be put out, another for the current to be turned on, and the operating work began.

I was struck by the instant change from nervousness to complete confidence which passed over Mr Marconi's face the moment his hand was on the transmitting apparatus – in this case a long wooden lever or key. He explained that it would first be necessary to transmit the letter 'S' in order to fix the attention of the operators at Poldhu, and enable them to adjust their instruments. This continued for a minute or more and then, with one hand on the paper from which he read and with the other on the instrument, the inventor began to send across the Atlantic a continuous sentence.

The opening salvo of crashing dots and dashes spelt out: 'Times London. Being present at transmission in Marconi's Canadian Station have honour send through Times inventor's first wireless transatlantic message of greeting to England and Italy. Parkin.'

There then followed short messages of greeting to Edward VII and Victor Emmanuel. Parkin wrote:

Outside there was no sign, of course, on the transverse wire from which the electric wave projected of what was going on, but inside the operating room the words seemed to be spelled out in short flashes of lightning. It was done slowly since there was no wish on this occasion to test the speed. But as it was done, one remembered with a feeling of awe, what he had been told – that only a ninetieth part of a second elapses from the moment when he sees the flash till the time when the record is at Poldhu.

During the hours of darkness in the days that followed more messages were sent and received, though there were periods when nothing would get through at all. For Marconi there were diplomatic niceties to be observed: he was Italian, but his company was English. From Edward VII he received the reply via Lord Knollys: 'I have had the honour of submitting your telegram to the King and am commanded to congratulate you sincerely from His Majesty on the successful issue of your endeavours to develop your most important invention. The King has been much interested by your experiments, as he remembers that the initial ones were commenced by you from the Royal yacht Osborne in 1898.' From Victor Emmanuel: 'The King learns with lively satisfaction of the splendid result constituting a new and glorious triumph for Italian science.' So much for Ambrose Fleming, George Kemp and a Marconi Company staff made up entirely of Englishmen. There was a ceremony with the officers from the *Carlo Alberto* in which both the British and the Italian flags were raised. Once again Marconi was fêted, with a banquet given in his honour in Sydney, Nova Scotia.

On 14 January 1903 Marconi travelled down to Cape Cod to supervise the first transmission from the United States to England. Four days later, President Theodore Roosevelt's message was flashed to Poldhu from the South Wellfleet station: 'To His Majesty King Edward Seventh. In taking advantage of the wonderful triumph of scientific research and ingenuity which has been achieved in perfecting the system of wireless telegraphy, I extend on behalf of the American people my most cordial greetings and good wishes to you and the people of the British Empire.' This had to be sent via the Glace Bay station, as Cape Cod could not transmit directly to Poldhu, and the reply from Edward VII had to be sent back across the Atlantic by cable, as Poldhu was not yet powerful enough to transmit to Cape Cod or Glace Bay. But the following day, 19 January, a message was received in Cornwall direct from Cape Cod, and for once Marconi appears to have gone almost mad with joy and relief.

One of the local men who had the job of keeping a horse and

buggy ready to take Marconi or his engineers to and from the Cape Cod station recalled: 'All of a sudden I see Marconi come tearing out of the plant with both hands full of white tape. I got my buggy all turned round and ready . . . When he come out again he had two big envelopes in his hand. They were messages to be telegraphed to Washington and New York. "Drive like the wind," said Marconi.'

The direct transmission from the United States to the western tip of Great Britain was an amazing achievement with the technology then available. But Morse sent by wireless remained incredibly ponderous, and was no match for cable in speed or accuracy. Richard Vyvyan had married just before the triumphant voyage of the *Philadelphia* in February, and his wife had been with him in America and at Glace Bay. On 3 January 1903 she gave birth to a daughter, and he sent a message to *The Times* in London: 'Jan. 3rd. Wife of R.N. Vyvyan – a daughter.' The letter 'E' in Morse is one dot, and atmospheric interference triggered the receiver's inker so that the message was recorded as beginning: 'Jane, 3rd Wife of R.N. Vyvyan'. There were many such mistakes that had to be corrected.

An intermittent news service was provided for *The Times* from 28 March. Then, on 6 April, disaster struck. What Nova Scotians called a 'silver thaw', a deluge not of snow but of freezing rain, weighed down the huge antennae with layer upon layer of dripping ice, until the entire structure collapsed. Marconi had overreached himself. There would be no Atlantic service for several years but Marconi's fame had already given rise to a ferment of speculation in the United States which the more sober commentators called 'wireless mania'.

23

A Real Colonel Sellers

When Frank Fayant of the American magazine *Success* came across Abraham White while investigating wireless fraud in 1907, he exclaimed: 'I have met a real Colonel Sellers in flesh and blood. Could I paint him with Mark Twain's pen!' In a retrospective thirty years later, the *Saturday Evening Post* sketched the unscrupulous wireless promoter.

> White's hair and mustache were flaming red; his eyes of china blue. He wore patent-leather shoes, a silk hat, a flower in his button-hole, a handsome gold watch-chain, a pear-shaped pearl scarfpin and a diamond ring that was not too big. He smoked corkscrew shaped cigars which he handed out freely, was never without a fat roll of $100 certificates [i.e. notes], which he peeled off with the easy indifference of an actor handling stage money.

To round off the picture, it transpired that White's real name was Schwarz – the German for black.

When railroad mania was at its height in the 1870s in the United States, Mark Twain created a character called Colonel Sellers. Sellers was a dreamer who imagined making his fortune from all kinds of unlikely enterprises, such as selling just one bottle of eyewash to every man and woman in China. A classic Sellers scheme was to promote a railway line to a new town he called

Napoleon. Everything was put in order for the arrival of the first hooting steam engine with its cowcatcher out front. Investors parted with their cash and waited for the grand opening, which they imagined would be a prelude to their making a fortune. On the day there was only one thing missing – the railway line. During the wireless boom there was no shortage of characters who in real life outdid the exploits of Twain's imaginary Colonel Sellers. To attract public attention and the funds of a credulous public, some built wireless stations, but these were usually so far apart that they were unable to exchange messages.

As a young man Abraham Black changed his name to Abraham White. He had moved from Texas to New York, having reputedly made $100,000 on an investment of forty-four cents, the cost of the postage stamp which had secured him shares which he sold at a profit before he had bought them. Stories about his audacious coups trading on the stock market were legendary. He was said to have made $25,000 while having a lunch which lasted under an hour. In December 1901 he had been inspired to try his hand as a wireless promoter by Marconi's success with the letter 'S' in Newfoundland. White had been impressed by the young American Lee de Forest and his fellow researchers in Chicago, who had formed the company which gave Marconi a run for his money with the 1901 America's Cup coverage. On 3 January 1902, White invited de Forest for lunch.

Ever since he was a boy, de Forest had dreamed of making his fortune as an inventor. Born in 1873, he had endured a lonely and difficult childhood. When he was eight years old his father, a Congregationalist minister, became head of Talladega College for Negroes in Alabama. The family were isolated from the white community, and Lee spent a great deal of time when he was not scrapping with local white children amusing himself with science experiments. His father wanted him to follow him into the ministry, but finally agreed to allow him to study at a college for poor children in Massachusetts, and from there Lee got a place in the scientific school at Yale. De Forest, like Reginald Fessenden, was a prolific inventor, and had notebooks full of what he regarded as brilliant ideas. He was most impressed by Marconi's achievements, took his doctoral thesis on Hertzian waves and saw his future as a wireless pioneer. Unlike Marconi, however, de Forest was always short of money, and when Abraham White invited him out for lunch and peeled off a $100 bill to 'get him started' he was wide-eyed with gratitude.

De Forest was just the man White was looking for: an ambitious, poverty-stricken inventor, with a modest track record and an over-riding desire for fame and wealth. Over lunch the De Forest Wireless Telegraphy Company was formed with capital of $3 million which appeared from nowhere. White became its President, and hired a press agent to fill the columns of the American dailies with glowing reports of de Forest's successes. De Forest went straight on the payroll at $20 a week, far more than he had ever earned before. A crucial plank of White's promotional campaign was to rubbish Marconi publicly and privately, calling him 'the Dago' and appealing to the nationalist sympathies of American investors. 'It is the policy of our company,' his advertisements would say, 'to develop its own system with American brains and American capital.'

De Forest became Abraham White's puppet. He was set up in a glass-sided laboratory on the roof of 17 State Street, Manhattan, and potential investors were invited to watch as he signalled to a station at the Castleton Hotel on Staten Island. In February 1903 a comical little motor vehicle with 'WIRELESS AUTO NO. 1' stencilled on the side could be seen parked in the Wall Street area, with White urging passers-by to observe de Forest sending stock market quotations to a nearby Dow Jones office. White had no compunction about inventing the most unlikely successes: to boost his company's share prices he would plant a story in the newspapers saying it had bought out American Marconi. Before there was time for a denial to be issued he would cash in shares whose value had risen momentarily, then keep his head down until the storm blew over.

White promised American investors that they would make fortunes in no time as he and de Forest established 'World-Wide Wireless', with stations all over the United States. He collected millions of dollars from investors, bought up other companies, and built wireless stations in many parts of America to maintain interest in his grand plan. An announcement would be made that the newly raised aerial would soon put the locals in touch with the rest of the continent; but few if any of these stations were able to contact each other, as they were too far apart for the makeshift technology de Forest was using. They were no more use than a Colonel Sellers railway station without the railway line.

While White was building his castles in the air and registering new companies with bewildering speed, de Forest was achieving in 1903 more or less what Marconi had three years earlier. The America's Cup was staged in October that year, with Sir Thomas Lipton again the challenger with his yacht *Shamrock III*. Once again Marconi and de Forest provided coverage for rival press associations, but this time an interloper spoiled the show. A Philadelphia offshoot of the fraudulent American Wireless Company simply jammed both Marconi and de Forest, sending streams of 'AAAA's and 'BBBB's and rude messages. De Forest, however, with White's help, managed to impress Sir Thomas Lipton himself, who gave him an endorsement and invited him to try his luck in England. White crossed the Atlantic and in 1904 began to set up wireless companies in London with industrial magnates at their heads.

In the United States, White founded *Wireless News*, a blatantly promotional paper which carried stories he commissioned himself. In an issue of April 1903 there was the following:

> Commercial wireless telegraphy, at a rate of one cent a
> word to the general public from Chicago to all principal
> points in the United States, will be an assured fact within
> ninety days, if the plans of the American De Forest Wire-
> less Telegraphy Company are carried out. Within sixty

days it will be possible to flash messages from Chicago to
steamers on the lakes, and to Detroit, Cleveland, Buffalo,
New York, and the Atlantic seaboard. Almost as soon, we
will be in wireless communication with St. Louis, Omaha,
Kansas City, and Fort Worth. A statement that these
things would be accomplished was given out yesterday at
the Chicago office of the company by Abraham White,
president of the corporation, and Dr. Lee De Forest, whose
inventions are claimed to have been made before those of
Signor Marconi.

White clearly had no intention of staying in the wireless business
long, but in the short term the fact that nobody quite knew what
to make of claims that wireless messages could be sent with ease
over thousands of miles, and that the cable companies were facing
serious competition, made life relatively easy for a trickster. He
indulged in the most outrageous scams. Not to be outdone by
Marconi's feat of sending messages of greeting from Canada and
America to European royalty, White despatched de Forest to Ireland so that he could receive from him what he liked to call an
'aerogram'. It was an eight-hundred-word history of telegraphy
which included the following: 'The application of man's genius and
the utilisation of God's natural forces represent a truly wonderful
combination ... this marvellous achievement recalls to mind that
historic telegraph message sent over the Morse cable many years
ago, "What hath God wrought!" '

As *Success* magazine put it when reviewing White's antics:

It would be unkind to suggest that the 800-word history
of wireless telegraphy, which White says he sent through
the ether to Glengariff Harbor, was in de Forest's pocket
before he set sail for Ireland. This great achievement in
aërography was recorded more than a year ago. Since then
nothing has been heard of the art in connection with
the de Forest companies, and it may be that transatlantic
aërography is one of the lost arts. It certainly does seem

strange to a layman that after sending an 800-word message across the Atlantic nothing more was heard of transatlantic messages. The cable companies still continue to
do business, and the owners of cable securities do not seem
to be lying awake nights worrying over aërograms.

Undaunted, de Forest and White continued to challenge Marconi with their publicity campaigns, and were soon to enjoy a most
unlikely scoop in the South China Seas.

Also snapping at Marconi's heels at this time was the man Lee
de Forest considered to be his greatest American rival, Reginald
Fessenden. In 1902 Fessenden had fallen out with his paymasters,
the US Weather Bureau. While Marconi was struggling to transmit readable signals from Glace Bay to Poldhu, Fessenden was
approached by two millionaires from Pittsburgh who said they
would back him and buy out his patents: Thomas H. Given, a
self-made man who had begun as an errand boy with the Farmers'
Deposit National Bank and risen to President; and Hay Walker
Jr, who had a company making soap and candles.

Given and Walker at least were genuine, and had no intention
of fooling investors with false promises of riches. Both put in their
own money to provide Fessenden with a salary and funds to turn
his experimental wireless system into a commercial and profitable
business. In November 1902 the National Electric Signalling Company was formed. Fessenden's first thought was to set up a wireless
link between Virginia and Bermuda, where he had taught a few
years before. But he discovered before this venture began that a
British cable company had a monopoly on communications in
Bermuda, and that he would end up in court if tried to break it.

Fessenden really had no idea how to turn his system into something saleable. Nor did his millionaire backers, who became more
and more frustrated with their chosen wireless wizard. Stations
were set up in Washington, Jersey City and Philadelphia, a few
signals were sent and attempts were made to interest the US Navy,
but nothing much happened. In frustration Given and Walker told

Fessenden that they wanted to compete directly with Marconi in establishing a transatlantic service. Fessenden was at first not at all enthusiastic, complaining that the greatest distance he had successfully transmitted a message so far was 120 miles. Fessenden was often moody and jealous, boastful one minute, despairing the next. But the money was there, and he took to the challenge his backers had set him with fervour. A station was established in 1904 at Brant Rock, just south of Plymouth in Massachusetts, and Fessenden moved there with his wife and son. A station on the British mainland would have to be found with the permission of the Postmaster General, who was now in charge of issuing wireless licences. As Marconi had always feared, America would very soon present him with fierce competition.

'In the public mind, Signor Marconi and wireless telegraphy are pretty nearly one; he is all of it. And for this there is some reason,' wrote the American magazine *Harper's Weekly* in February 1903. 'Marconi was the first in the field, the first to send a wireless message several miles, the first to reach a hundred miles, and the first to cross the sea. He has had the lead, and he has it now. And this, in the face of a perfect host of competitors, is a big achievement for a young man still under thirty. He deserves all the fame he has won.' But, *Harper's* warned, the wireless maestro was about to be put in his place by two inventors working in the United States: Lee de Forest and Reginald Fessenden. Taking at face value the outlandish claims of de Forest's unscrupulous backer Abraham White, and Fessenden's boasts about the superior technology he had developed, *Harper's* foresaw the end of Marconi's reign as the leading exponent of wireless.

The pace of Marconi's life now was such that he barely had time to consider the threat of rival technologies or the American competition. Take just one year, 1903. In January he was working on the first transatlantic transmissions from Cape Cod via Glace Bay. In May he was invited to Italy to be made a Citizen of Rome and visited his home town of Bologna on the way, to a rapturous reception. At the railway station in Rome he had to fight through crowds of admirers to reach the Mayor's coach. A group of students unhitched the horses and pulled the coach to the Grand Hotel. Marconi's parents were with him: this was one of the last times he saw his ageing father, who was nearly eighty and very proud of his famous son. The German Kaiser was visiting the Vatican, and the streets were so jammed that Marconi was an hour and a half late for a lecture he was giving in Rome. Italian newspapers had a bit of fun about the antagonism of the two 'Williams', Guglielmo and Wilhelm. That night Marconi was introduced to the Kaiser by King Victor Emmanuel. It was an awkward occasion in which the German ruler expressed his country's disapproval of Marconi's refusal to communicate with German wireless services.

In July the Prince and Princess of Wales paid an official visit to the wireless station at Poldhu, arriving in a motor-car, one of the first to be seen in that part of the world. The Poldhu Hotel was decked with patriotic flags, as were the masts of the station. Marconi was the star of the show, demonstrating to the future King George V, who could read Morse code, the reception of signals from the nearby Lizard station. A small party including the Prince climbed to the top of one of the four transmission towers to take in the view. Then, in a cloud of dust, the royal party motored over dirt roads to the Lizard, where they enjoyed a brisk clifftop walk before taking tea with Marconi at the Housel Bay Hotel.

In August Luigi Solari, the Marconi Company representative at a Berlin conference called by the Germans, stormed out in protest at attempts by the host country to get an international agreement about communication between rival systems of wireless. The Kaiser, still fuming about the affair of the *Deutschland*, wanted to bring diplomatic pressure to bear on Marconi to force him to tune in to German wireless. Brushing this challenge aside, Marconi and Solari fitted out the Cunard liner *Lucania* with wireless and sailed for New York on 28 August, demonstrating to passengers and to officers from both the British and Italian navies that he could receive signals either from Poldhu or Glace Bay all the way across the Atlantic. With the help of these signals he began publication of the first ever regular mid-Atlantic daily newspaper.

While they were in New York Marconi and Solari paid a visit to Thomas Edison, and were amused when he found he had nothing to offer them for lunch: his wife was away and the cupboard was bare. Edison did, however, make them a present – he transferred some of his patents to Marconi. These were of no practical value except as a safeguard against possible scurrilous litigation in America. In October Marconi was again aboard the *Lucania*, steaming for Southampton, all the while testing his equipment. On this journey he enjoyed another shipboard romance, this time with a very remarkable young woman. Inez Milholland was easily a match for the famous Marconi, despite the fact that she was barely eighteen years old. The daughter of a *New York Tribune* reporter who had made a fortune from the invention of a pneumatic tube to send messages around offices and stores, Inez was well educated, an amateur actress and a fine athlete – at Vassar College she had set records for the shotput and basketball throw. Her frustration at being barred from male colleges and her concern for the rights of the underdog led her to become a staunch suffragette. As Marconi's daughter Degna would remark, Inez was not really her father's type. Nevertheless, by the time the *Lucania* docked they were engaged.

At the end of this momentous year of 1903, Marconi was back at Table Head on Cape Breton. In November he visited Richard Vyvyan's wife in hospital, and took a fancy to the matron. When Cuthbert Hall, a Marconi Company director back in London, heard that Marconi had shown the matron round the Glace Bay station he sent a stern rebuke about the need to maintain security, accusing the man who had made wireless history of being the 'chief sinner' when it came to breaking the rules about keeping out prying eyes. At this point in Marconi's life it is almost impossible to keep up with him. His companies were operating around the world, while he barely paused except for the occasional few days at the Haven Hotel in Poole.

He had not long returned from America, again on the *Lucania*, when he joined his old British Navy friend Captain Henry Jackson on HMS *Duncan* en route to Gibraltar. Jackson had risen through the ranks, and now had command of his own ship, which he put at Marconi's disposal to test the possibilities of communication between Poldhu and the Mediterranean. The Marconi Company was about to sign a long-term agreement with the Royal Navy to supply them with equipment and expertise, a contract that was to run until 1914. When he left Jackson Marconi went to Italy to set up a wireless service with the Balkans, and to discuss the building of a powerful station at Coltano, near Pisa. And then there was a new station in Scotland to test overland transmissions from Poldhu.

It is hardly surprising that, with all his other activities, at the end of 1903 Marconi was pipped at the post in a novel use of wireless. Marconi engineers had taken equipment to the Boer War in 1900, but it had not worked well and was not used by journalists. Soon a conflict on the far side of the world would offer the first real opportunity for wireless to prove its value to war correspondents as well as to the belligerent armies and navies. Of the companies then operating, Marconi's was far and away the most able to offer expertise and equipment to journalists. But the closed social world of a transatlantic liner robbed Marconi of the opportunity to enjoy another historic first. Instead it fell into the lap of his American rival Lee de Forest.

24
❖❖❖❖❖

Defeat in the Yellow Sea

Late in December 1903 the White Star Line's *Majestic*, which had recently been refitted after serving as a troopship in the Boer War, eased out of Liverpool docks en route for New York. Among the first-class passengers were the two men who regarded themselves as the foremost exponents of wireless telegraphy in the United States, Professor Reginald Fessenden and Lee de Forest. They were not on speaking terms, as Fessenden was suing de Forest for infringement of his patent on the new electrolytic detector, or barreter. Both were returning home after exploring the possibility of establishing their wireless systems in Britain. Fessenden had been looking for a site to set up a station for his transatlantic venture, while de Forest had been taking part in a competitive demonstration of wireless for the General Post Office. Neither man was in a very happy mood: Fessenden was being offered a remote site in Scotland by the Post Office, and de Forest felt he had been rejected by British 'hauteur', as he had received no orders for his equipment.

However, de Forest's mood changed when he met aboard ship a British war correspondent, Captain Lionel James, who was on the first leg of a journey out to the Far East, where war was brewing between Russia and Japan over disputed territories in Manchuria and Korea. Captain James was a classic product of the British Raj, a former tea-planter and racehorse-owner. Four years earlier he had taken a gamble by offering himself as a special war correspondent for *The Times*. He had reported on the Boer War and had been with General Kitchener in the Sudan, where he had made a name for himself as a front-line trooper in the world's press corps.

Though both de Forest and James liked to claim credit for having the idea of using wireless to report the Russo–Japanese War, it is most likely the plot was hatched by chance on the *Majestic* when the two met and fell into conversation. James had heard of de Forest that October when he had been in New York, and had been given the impression, probably by Abraham White, that the American was way out in front when it came to wireless telegraphy. De Forest, in his autobiography, claims to have heard of James, and to have persuaded him to use his wireless in the Far East. On the voyage to New York it was agreed that if James could persuade *The Times* to put up the money, de Forest would provide wireless equipment and engineers to establish a station somewhere on the China coast. James anticipated that if war broke out much of the naval action would take place in the Yellow Sea, which is enclosed by the coastlines of Manchuria, Korea and China. To the north was the Russian naval base of Port Arthur. If he could charter a reasonably swift steamer and get permission from the Japanese navy to cruise among its ships, there was a good chance that with de Forest's wireless equipment he could beat by hours – even days – his fellow correspondents, who would have to find cable stations from which to send their reports.

From New York James cabled London, and a reluctant *Times* accountant agreed to pay £1000 for de Forest's men and equipment, and to put up a further sum for the hire of a suitable steamer. James gave the responsibility of finding the site for a shore station to a junior reporter, David Fraser, who had travelled with him on the *Majestic* as his 'valet', a ruse to conceal from other journalists the purpose of their voyage. James and Fraser took the train across America and shipped from San Francisco on the *Siberia*, hoping that they would reach the Yellow Sea before war broke out. As it was, they saw no signs of any action as they approached Yokohama in Japan at the end of January 1904.

Once James had secured the agreement of *The Times* to pay for all the necessary equipment and a ship, de Forest was in a quandary. The only transmitters and receivers he had were in Ireland, and he had to beg and bribe his motley crew of helpers to pack them up and ship them to New York. He then had to find volunteers to take them out to the war zone. Two of these, 'Pop' Athern and Harry Brown, caught the train to Seattle, from where they boarded the last ship to Yokohama that would give them time to set up a simple shore transmitter and fit out whatever steamer James had been able to lay his hands on. From that point onward, de Forest could only wait and hope that all went well, and that for once he might really make history and attract the admiration and credit he craved.

While Captain James negotiated with the Japanese authorities in Yokohama for permission to cover any future conflict, David Fraser was instructed to stay aboard the *Siberia*. In his pocket he had a telegram which read: 'Shantung Peninsula best erect mast 180 feet high 30 feet from water's edge. De Forest.' Though he could see on his map that the Shantung Peninsula was on the eastern coast of China, in a very favourable position for scouting the Yellow Sea, Fraser was otherwise very much in the dark. 'Of wireless telegraphy and all that pertained to it I was completely ignorant,' he wrote in *A Modern Campaign*, an account of his adventures. 'My orders were to proceed to Shantung and there prepare for the plant and operators which were coming from America. Where to establish the station I had to decide; and how to place so mighty a mast, the cable told. But how to get the thing to stand up, how to procure the materials, where to find the lunatics mad enough to climb the mast when it was up, were problems which refused to solve themselves.'

It was with some apprehension that Fraser approached Shantung, which to him 'suggested missionary-eating natives and other vague horrors. The little red dot on the map, not far from the

Promontory, was obviously the place upon which to base operations.' This 'red dot' was a far-flung and isolated outpost of the British Empire, the naval base of Wei-hai-Wei, a kind of northern Hong Kong leased from the Chinese. Here Fraser found his own people, and an engineer to supervise the erection of the mast. He was loaned a horse to search for a suitable site. The Chinese would not allow him to use their territory, so a place had to be found on the bleak coast of little Wei-hai-Wei itself, far from the conviviality of the British clubhouse and expatriate community. Before he had settled on a suitable spot, Fraser received a cable from James which read: 'Expedite forestry scrap imminent.' In other words, 'Get the de Forest station set up: the war is about to begin.' In fact the conflict had already begun by then, without either side formally declaring it. Two days after Fraser had arrived in Wei-hai-Wei, on 8 February 1904, Japanese torpedoes had sunk three Russian battleships as they lay in the harbour of Port Arthur.

The engineer Fraser had seconded to him, a man called Griffin, solved the problem of materials for the great aerial pole by buying old masts from Japanese junks. Permission to put it up was granted through the old-boy network, as one of the British officials in Wei-hai-Wei had known Fraser's brother Tim out in India. Fraser received another telegram from James, which said he had chartered a steamer called the *Haimun*, which would be arriving in about ten days' time. He repeated the command 'expedite forestry'.

Raising the mast involved the labour of 'fifty Chinamen', according to Fraser, plus a hundred naval ratings he had managed to borrow from a sympathetic British officer. After ten days of hauling and heaving it was almost up when it snapped, throwing 150 men this way and that like defeated contestants in a tug-of-war. The mast was still not in place when the *Haimun* hove to. The ship's regular crew had disappeared when they learned the dangerous nature of the voyage, and had been replaced by a makeshift assembly of 'rickshaw drivers and coolies', according to Fraser. However, an experienced captain had remained in command, and there were six British officers aboard, as well as a lady interpreter.

When the other war correspondents learned what James was up to they set up a chorus of protest. The British Minister in Tokyo, Sir Claud Macdonald, told James he was wasting his time with wireless. Any attempt to get permission from Japan to steam among its battleships would be a flagrant breach of neutrality, and James was warned by a British admiral that the Japanese navy would sink the *Haimun*. But for James headlines were more important than international diplomacy, and he came to an agreement with the Minister of State for the Japanese navy that the *Haimun* would offer naval intelligence exclusively to them in return for the freedom to send reports back to *The Times* in London. The fact that he had offered to spy on the Russians was only revealed twenty-five years later in his memoirs. The Wei-hai-Wei station and the receiver on the *Haimun* picked up both Russian and Japanese wireless signals, and James fulfilled his bargain with Japan by reporting the position of a Russian station, which was quickly attacked and put out of action.

James was able to report from the Yellow Sea between March and June 1904, and to outdo his rivals with his first-hand accounts of naval manoeuvres. Inadvertently, the very British Captain James and *The Times* had given Marconi's American rival de Forest the publicity he so desperately wanted.

When the Marconi Company got wind of James's plan to use de Forest wireless in the Far East they tried to beat him to it. Marconi and Solari sounded out the Italian navy about the possibility of a ship to take equipment and operators to the Yellow Sea, and Alfred Harmsworth, proprietor of the *Daily Mail*, offered to put up the money if they thought they could beat de Forest. But they were too late. To rub salt into their wounds, Captain James effectively shanghaied the *Daily Mail* correspondent, F.A. McKenzie, whom he found under Japanese house arrest in the port of Chinampo. McKenzie had been with the Japanese troops when they had first encountered the Russian Cossacks at a battle on the Manchuria–Korea border in May 1904, and he told James he had 'the story of the war'. Now he was stranded, unable to get to a cable station and forbidden by the Japanese to file any stories at all.

On the pretext of helping McKenzie, James took him aboard the *Haimun* and promised to do what he could to find him a cable station in China. Instead, while sending his own wireless reports, he made sure the *Daily Mail* man remained out of action. As the steamer approached Wei-hai-Wei, James held a dinner party. He recalled gleefully some years later: 'We plied McKenzie with as much champagne as he would take, and pointed the crevices with cocktails and liqueurs. McKenzie deserved these good things, as he had been living on Japanese dried skate and sodden rice for the past fortnight. The comfort of the *Haimun* was a joy to him.' Once McKenzie was asleep he was locked in his cabin. James took a boat ashore to Wei-hai-Wei, checked that his stories had got through, and returned to the *Haimun*. As he put it, 'as in love, so in war – and especially War Correspondence – all is fair'.

The Russians soon caught up with the *Haimun*, however, and life became more difficult for James. Fraser was sent off to cover the land battles in Manchuria, and the Japanese withdrew James's permit, bringing to an end his short, heroic and duplicitous engagement as the first war correspondent to make use of wireless. The Wei-hai-Wei station was dismantled long before the decisive engagement between Russia and Japan the following year, but back in the United States de Forest and Abraham White made much of their brilliant coup.

And Guglielmo Marconi had had to endure in 1903 another humiliation, this time in his own backyard, in one of the most august of London's scientific institutions.

25

A Wireless Rat

Demonstrations of scientific wonders were still very popular with the London public in the early years of the twentieth century, and there was no lecture theatre more prestigious than that of the porticoed Royal Institution in Albemarle Street, off Piccadilly. Here the likes of Marconi's boyhood hero Michael Faraday had drawn huge crowds in the nineteenth century, and in 1900 Marconi himself had taken the platform to demonstrate the miniature coherer he had devised. The Royal Institution's audiences were described by *The Electrician* as a mixture of those who regretted they had not had time to read up on the subject beforehand in the *Encyclopaedia Britannica*, scientists hoping to hear of a new discovery, and 'a large proportion of the fairer sex, a number of whom were old *habitués*'.

Ambrose Fleming was a popular speaker here, and in the early evening of 3 June 1903 he prepared for a lecture on the progress made in wireless telegraphy, with special reference to the ability to tune to specific wavelengths. As a demonstration, messages relayed from Poldhu via the Marconi station at Chelmsford in Essex would be received at prearranged times on a tickertape machine on the platform. Fleming was warming to his subject of 'Electric Resonance and Wireless Telegraphy' when his assistants noticed something strange happening. One of them, Arthur Blok, recalled:

One of the Marconi Company's staff was waiting at the Morse printer, and while I busied myself with demonstrating the various experiments I heard an orderly ticking in the arc lamp of the noble brass projection lantern which used to dominate this theatre like a brazen lighthouse. It was clear that signals were being picked up by the arc and we assumed that the men at Chelmsford were doing some last minute tuning up. But when I plainly heard the astounding word 'rats' spelt out in Morse the matter took on a new aspect. And when this irrelevant word was repeated, suspicion gave place to fear.

Fleming's poor hearing protected him from any distraction and he lectured on oblivious as Blok and the other assistants heard in Morse: 'There was a young fellow of Italy, Who diddled the public prettily.' Then there were some quotes from Shakespeare. All this was recorded on a tape which was spewing out before any planned signals were due. As the time for the Chelmsford message approached, Blok looked anxiously around the audience to see if there were any telegraphists who had been able to decipher these unwanted messages. Everyone, it seemed, was enthralled by Fleming, and nobody had noticed anything untoward. Then Blok's eyes lighted on a face he recognised: a young man associated with Nevil Maskelyne, son of a celebrated magician who performed at the Egyptian Hall around the corner,* and an aspiring rival of Marconi.

Fleming concluded the lecture without any knowledge of what had happened backstage. The audience applauded, and the event was written up as a great success in the technical magazines and newspapers. Fleming had a reputation among his students at University College for being quick-tempered, and it seems nobody

---

* Built in 1812 to display arts and curiosities from around the world, the Egyptian Hall later became a place of entertainment. The American showman Phineas T. Barnum exhibited the dwarf Tom Thumb there in 1844, and Nevil Maskelyne performed his magic acts in the years before it was demolished in 1905.

dared to tell him of the 'rats' intrusion until a day or two later. When Blok showed him the offending Morse tape, which he had quickly gathered up and stuffed in his pocket, the distinguished scientist flew into a rage.

As no publicity had been given to this attempt to discredit Marconi by showing that his transmissions were vulnerable to attack, it could have been quietly brushed aside. But Fleming could not contain his anger. He sent a letter to *The Times* which described the act as 'scientific hooliganism', and lashed out at 'monkeyish pranks' – phrases which instantly became popular amongst his students. Fleming appealed for anyone who knew who had committed this 'outrage against the traditions of the Royal Institution' to come forward.

As it happened, the culprit came forward himself, with the telling argument that he was merely illustrating the vulnerability of wireless signals to interference, and making nonsense of the claim that tuning would provide confidentiality. In a letter to *The Times* published on 16 June, Nevil Maskelyne owned up without apology, rejecting the charge of 'scientific hooliganism': 'Prof. Fleming says his experiments were carried out in the face of a deliberate attempt to wreck his demonstration. That is a falsehood. His demonstration succeeded by courtesy of those who, having the power to wreck it, yet refrained from doing so.' Far from having carried out a 'monkeyish prank', Maskelyne declared that he and his accomplices were making a valid point: all the claims made for the 'syntony' or tuning of wireless waves emanating from Poldhu were false, and the public ought to know it. 'We have been led to believe that Marconi messages are proof against interference. The recent Marconi "triumphs" have all been in that direction. Prof. Fleming himself has vouched for the reliability and efficacy of the Marconi syntony. The object of his lecture was to demonstrate this. Then, if we are expected to believe certain statements, no one can complain if we proceed to put those statements to the test.'

Evidently relishing the discomfort he had caused Fleming, Maskelyne told newspaper reporters that he had decided not to wreck the lecture, but merely to draw the fire of a pompous academic making false claims. He informed *The Electrician*: 'This device succeeded perfectly. The simple interjection of the word "rats" actually drew Prof. Fleming himself. It is a harmless expression signifying incredulity. I have heard it used by university professors on occasions; and really, it was most appropriate. As for instance when the lecturer spoke of signals from Poldhu travelling over vast areas without anyone therein having cognisance of the fact. Surely he knows that a "sweet little cherub" in the form of a "Maskelyne" receiver generally keeps watch upon the doings at Poldhu.' It was true that Marconi had admitted in a newspaper interview that Poldhu had been 'tapped' by a pirate station in Cornwall.

Maskelyne had Fleming and the Marconi Company over a barrel. In a series of interviews with the *Express*, *Telegraph*, *St James's Gazette* and *The Times*, Fleming tried desperately to rubbish Maskelyne, claiming that the receiving instrument at the lecture was not syntonic, that it was 'demonstration' rather than working equipment, that the interference had stopped because his assistants were able to cut it out, that it only happened when they were tuning in, and that it was not wireless interference at all, but used 'earth currents'. As Maskelyne gleefully pointed out, the eminent Professor contradicted himself almost every time he sounded off on the subject. The row rumbled on in the newspapers, with headlines such as 'Wireless "Rats" – Mr Nevil Maskelyne and his Mysterious Message' (*Morning Leader*), giving the opportunity for a very public discussion of the true state of the Marconi wireless system. Maskelyne pointed out that his surreptitious monitoring of Poldhu, which he called 'the thunder factory', had revealed that the fifty-horsepower transmitter sent out messages at two and a half words per minute. He told the *Morning Leader*: 'Had the same horse power been applied to a motor-car the message would have arrived as quickly by road.'

There were suggestions in the press that a cable company intent on sabotaging wireless was behind the Maskelyne scam. That is possible, but there was never any hard evidence. Maskelyne was most likely just another jealous rival who wanted to cut Marconi down to size. In 1904 he was briefly involved with Abraham White when the American huckster tried to set up in competition with Marconi in London. Whatever his true motives, Maskelyne's criticisms were valid. Morse by wireless was still agonisingly slow, and it could not promise privacy. But if the Marconi Company was being a little careless of the truth, the claims it made for a wonderful future for wireless were nothing compared with the hyperbole being orchestrated by its rivals on the other side of the Atlantic.

26

Dazzling the Millions

Among the wonders that greeted the millions of visitors to the
World's Fair in St Louis which opened on 30 April 1904 was
a tower which had written on it vertically in illuminated letters,
each eight feet high, the name 'DE FOREST'. A report in the
St Louis *Post Dispatch* painted the picture:

> Flashing messages through space from the Fair to the
> office of the *Post Dispatch* continues to be the wonder of
> Fair visitors and crowds watch the process from morning
> until night. The flash of 20,000 volts every time the opera-
> tor presses his key is to them a thing of fascination. Then
> they turn from it to look from the great De Forest tower
> out eastward across the large city, but they see no sign of
> the message which the clicking instrument is sending out
> there through space. Sometimes they stop the operator at
> his work to ask him if it is really so.
>
> They shake their heads in amazement when he answers
> yes, and explains that in the *Post Dispatch* office another
> instrument is ticking in response to his, and thus carrying
> Fair news to the newspaper and the world. The loud buzz-
> ing of the powerful instrument surrounding the operator
> 200 feet above the ground in the De Forest tower does

not prevent the visitors from crowding about him. It is so loud that the operator must keep his ears full of cotton. It fairly deafens visitors and sending [sic] them away with a headache if they stay too long, but nevertheless they stay, for the power of the mystery is very great ... The dots and dashes are so audible that operators for telegraph companies and the police and fire departments anywhere within two blocks of the wireless tower amuse themselves with reading the wireless messages as they are buzzed off by the sending operator.

The purple prose reads like a press release, and it almost certainly was, the work of the redoubtable Abraham White, Lee de Forest's bejewelled and moustachioed backer, whose wireless companies still had little or nothing to show for the millions of dollars invested in them.

When the World's Fair was being planned to commemorate the centenary of Napoleon's sale of Louisiana to the United States, the Marconi Company had been invited to show the American public, most of whom had only the vaguest idea about wireless telegraphy, how this amazing new invention worked. However, Abraham White put in a successful bid to demonstrate his 'All American' system, and the Marconi Company pulled out when it was understood that there would be competition. It was not their style to engage in that kind of rivalrous showmanship. This left the field open for White and de Forest to pull out all the stops and to lay claim to an array of innovations which by that time were in reality already part of wireless history. With his unerring eye for the main chance, White bought an old observation tower which had been used to give the public an aerial view of Niagara Falls. The tower had been dismantled because of the danger of icicles falling from it onto the people below, and White had it shipped to St Louis and reassembled as a primitive wireless transmitter at the World's Fair, with de Forest's name emblazoned on it. The company's role in the pioneer reporting of the Russo–Japanese war by wireless was celebrated in a large map of the Yellow Sea within the de Forest showcase.

The impression given was that the de Forest Company was well established and fully staffed. For a few weeks de Forest was set up as a gentleman with a butler and his own carriage. This was no more than a façade. Although he was raising huge amounts of capital, Abraham White allowed de Forest no more than a pittance for research: at the World's Fair he began with only one assistant. When a young telegraph operator, Frank Butler, turned up to offer his services he was taken on only after agreeing that he and the other assistant would share one salary. None of this, however, was evident to the public, as 'aerograms', as White still liked to call them, were transmitted to local newspapers, and a bulletin was produced on site which made extravagant claims for de Forest's system.

Marconi himself paid a brief visit to the St Louis World's Fair, where he drank too many mint juleps and became unsteady on his feet. It was as if all his pioneering work in wireless, his transmissions over thousands of miles, his shipboard news bulletins on the Cunard liners, had never happened. Visitors to the Fair were led to believe that this wonderful new system of communication was being demonstrated for the first time. De Forest's assistant Frank Butler recalled for *Radio Broadcast* magazine in 1924:

> At night the tower was illuminated by thousands of electric lights which could be seen for many miles. In addition to this station, another exhibit was maintained in the Electricity Building and from both places we demonstrated 'wireless' to endless streams of curious people. In an adjoining booth was displayed 'Wireless Auto No. 1,' which was the very first wireless automobile. Its range of reception was only a few blocks but it always created much interest whenever it was driven about the streets or viewed at its exhibitor's stand.

Encouraged by the popular interest in wireless, de Forest decided to build a larger transmitter, nothing like the size of Marconi's new station on Cape Breton, but much bigger than anything he himself had attempted before, to send a signal from St Louis to Chicago, three hundred miles away. Huge, unstable batteries were cobbled together on the spot to generate giant sparks which sent thunderclaps across the exhibition grounds. A kind of pump handle had to be fashioned to fire the signals. Butler said tapping out messages on it was like 'working at the village well for half an hour at a time'.

De Forest was an impatient and shoddy worker, and often his batteries would blow up. The air was filled with static electricity which fired bolts at Butler's head. He recalled: 'The roar from the spark gap could be heard a block away and it held its own in noise intensity with the ballyhoo bagpipe of the Jerusalem Exhibit on the one side and the cannonading in the Boer War Exhibit on the other. The odor of ozone, mixed with kerosene, was always present.' They finally got the transmitter to work in September, and on 'Electricity Day' at the Fair de Forest managed to send a signal to the Railway Exchange Building in Chicago. Judges were stationed in Chicago and St Louis to ensure there was no cheating. De Forest claimed a record distance for wireless telegraphy over land, and was awarded the Grand Prize.

Abraham White's publicity campaign brought orders for equipment from the Boston-based United Fruit Company, whose fleet of ships carried South American produce to the United States. As with all de Forest equipment at this time the system delivered to United Fruit was crude, and worked only erratically in the highly static atmosphere of the southern seas. But the company so valued the novel ability to keep in touch with its cargo vessels that it established shore stations in Costa Rica, Panama, Nicaragua, Cuba, Louisiana and some West Indian islands, and fitted all its ships with wireless.

The United States Navy too responded to White's hyperbole, asking de Forest to set up stations in the West Indies. For three years the naval authorities in America had recognised the potential of wireless, but had been unable to make a decision about which was the best equipment to buy. They were not helped by their chosen representative in Europe, Commander Francis M. Barber, who in 1901 had been brought out of retirement to advise them on the relative merits of rival inventors. Commander Barber lived in Paris, spoke fluent French and German, and moved in elevated diplomatic circles. Much depended on his findings, but he disliked inventors as a species, feeling they were always making false claims.

Barber's official reports to his old friend Admiral Bradford, chief of the Bureau of Equipment, who had initially favoured the Marconi Company, are peppered with dismissive accounts of most of the European contenders in the race to dominate the wireless industry. Count von Arco of the German Slaby-Arco system he described as 'a weedy little chap with a great big head – he looks like a tadpole'. Barber went to and fro between two French rivals, Ducretet and Rochefort, who in turn insulted each other. But the man he had most contempt for was Marconi. All the criticisms thrown at 'the inventor of radio' by Oliver Lodge and, latterly, William Preece, were relayed across the Atlantic with a kind of wicked glee. Marconi had stolen everyone else's ideas, and he was bound to fail. It was highly unlikely that he had really transmitted across the Atlantic. Barber went so far as to quote Colonel Hozier, the Secretary of Lloyd's of London, who were negotiating a new contract with the Marconi Company (of which Hozier was also a director): 'He thinks Marconi had never yet got a signal across the Atlantic or 2000 miles at sea either. The whole thing was a stock-jobbing operation worked in the interest of "a lot of Jews".' Barber wanted the American Marconi Company driven out of business.

Between 1901 and 1904 the US Navy tested every available wireless system, and began lobbying for this new technology to be brought under some kind of government control. At international conferences it began to side with Germany in its 'malignant Marconiphobia'. The American Marconi Company fought a determined rearguard action opposing government regulation. Wireless telegraphy, which just three or four years earlier had been so novel, magical and exciting, was now becoming tangled up in bureaucracy. For Marconi himself there were no new 'great things' to be achieved. At Glace Bay and Poldhu his engineers were no nearer creating a regular transatlantic service, and their failure was undermining the value of the company's shares. A new and larger transmitter was being built at Louisberg on Cape Breton at tremendous cost. The weight of responsibility on Marconi's shoulders was almost insupportable, but he continued to live a life of near-perpetual travel. He only rarely visited the Villa Griffone, he had no permanent home in England, and he rarely saw his fiancée, the beautiful firebrand suffragette Inez Milholland.

27

'Marky' and his Motor

The one place Marconi could return to from time to time to relax was the Haven Hotel in Poole. Here he was wined and dined by distinguished visitors to the south coast resorts, and he became a frequent and much-admired guest of a wealthy Dutch couple, Charles and Florence van Raalte. In 1901 they had bought the whole of Brownsea Island, which lies just off Poole harbour. Over the centuries the island had had many owners, most of whom lost money trying to exploit it in one way or another. The van Raaltes, however, acquired it purely as a private playground for their own pleasure and amusement, and to entertain their upper-crust friends and celebrities of the day. When he was staying at the Haven Hotel, Marconi often made the short crossing to Brownsea to be entertained by the van Raaltes in their newly refurbished castle, and he became a favourite guest with their teen-age daughters. The place may have reminded him of the Villa Griffone, where he spent so much of his childhood with his English cousins.

One of the van Raaltes' daughters, Margarite, recalled years later the delights of the Haven Hotel and how enchanted Marconi was with Brownsea Island when he was invited to stay.

The Haven Hotel opposite was splendidly run by a Frenchman and his wife by the name of Poulain, whose daughter had yellow tassels to her high-laced French boots. The cooking was most excellent: and here, rooms and a big workshop were permanently kept by Marconi, the inventor of wireless telegraphy ... My brother, a born mechanic and very much of his generation, was madly interested and I fascinated, by discovery after discovery and at the development of each new invention.

All the van Raaltes' guests were given nicknames such as 'Poops' or 'Winkle', and before long Marconi was referred to affectionately as 'Marky'.

Marky became a great friend of ours and came to Brownsea whenever he cared to: we also, Nony [her brother] and I used to row over and call on Mr Kemp his chief engineer and right-hand man in the workshop, and ask what was going on ... Frankly I did not understand much about it, and was greatly impressed when one Christmas Marconi gave us a set consisting of a small wooden box about a foot square by six inches deep with an aerial three or four feet high and a tickety tick apparatus on the box. It looked so simple, but would not work. Then Nony and Tommy mastered it; and we set it up between the bedroom in the tower and Nony's workshop in the Villino [a small house beyond the walled front garden where Nony worked]. We 'Morsed' messages to each other and got so keen on the Morse code, that we could 'left eye right eye' across the luncheon table!

In the relaxed atmosphere of Brownsea Island, with its pine woods and lakes, Marconi, a celebrity who was amusing, modest and mysterious, made a very favourable impression on the van Raalte girls. 'Marconi was easy and pleasant,' Margarite recalled. 'He absolutely could not explain his inventions though he was clever enough to think them up.'

Much as he enjoyed Poole and Brownsea Island, Marconi was

able to make only fleeting visits. In March 1904, when Lee de Forest's wireless operators were making history of a kind in the Yellow Sea, Marconi and a team of engineers were in Italy, where the company had been asked to set up what would be the largest transmitting station in the world at Coltano, a village on flat marshland near Pisa. Marconi spent much of his time in Bologna, where his ageing father was seriously ill and the family had gathered, fearing that he would not live much longer. On 25 March Giuseppe's condition suddenly deteriorated, and in the early hours of the following day he died. Guglielmo, with his hectic schedule, just had time to attend the funeral a few days later. In May Marconi was in Italy again to receive an Order of Merit from the King: if his star had fallen in America, Europe still showered him with honours, and he had now acquired the lifestyle of the most favoured sons of the upper classes and aristocracy. This included an enthusiasm for the brand-new and dreadfully hazardous sport of motoring.

As early as the summer of 1903 Marconi had arrived at the Poldhu Hotel not on his motorbike but in a spluttering four-wheeled vehicle which raised a cloud of dust on the country roads. George Kemp recorded in his diary for 30 July: 'I went to the Lizard Station in the afternoon and on to St Kevern with Mr Marconi in his new Napier car.' Ominously, a brief entry for the following day reads: 'The motor car I saw in pieces in the shed. In the afternoon Mr Marconi went to Gunwallow Church with me.' Had Marconi crashed? It was quite likely, as these were very early days for motoring in England, and Marconi had joined an elite band of aristocrats and millionaires who were the first to threaten life and livestock on the unmetalled roads. In December 1903 Kemp noted that Marconi had arrived in 'the Car' with two drivers. There were no driving tests then, and the pioneer motorists were notorious for their flouting of the laws that Parliament quickly enacted in an attempt to control the wonderful but often fatal freedom of the open road.

The invention of a practical, petrol-driven motor-car had more or less kept pace with Marconi's own development of wireless telegraphy. All the early models were put together in France and Germany, where there were no legal restrictions on speed. In November 1896 a twelve-mile-per-hour limit was introduced in England, celebrated with the first London-to-Brighton 'rally'. A collection of thirty-three self-propelled vehicles, including tricycles, attempted the sixty-mile run from the capital to the coast. Of these, fourteen made it to the finishing line, most of them foreign-made vehicles. Participants were warned: 'Owners and drivers should remember that motor cars are on trial in England and that any rashness or carelessness might injure the industry in this country.'

The racing of cars had begun in privately organised events, and in 1900 the *New York Herald*'s flamboyant owner Gordon Bennett Jr had put up a trophy, the Coupe Internationale, for the winner of a race from Paris to Bordeaux. There was so much death and destruction on the 1903 run that stricter regulations were introduced. But by then the number of car owners had begun to grow rapidly both in Europe and in Britain, and new laws were created to keep them in check. In England the Motor Car Act, which came into force in January 1904, set a new upper speed limit of twenty miles per hour. By the end of that year there were 8500 motorists on the roads, one of them Marconi, who bought a white Mercedes in which he would drive from London down to Poole. Built by the German Daimler company, the 1904 Mercedes had proved itself the most advanced and fastest automobile on the roads. Sometimes Marconi raced wealthy aristocratic friends like Howard de Walden, whom he met at the van Raaltes' on Brownsea.

Marconi in his Mercedes had to be ever-watchful for 'trapping', which became something of a sport for local police, who hid by the roadside with their stopwatches at the ready, in wait for speeding motorists. The motoring lobby argued that this was not a proper part of police duty, and that the law was unenforceable because of 'inaccurate stop watches'. But a total of 1500 motorists were fined in 1904–05 for driving 'furiously, negligently, recklessly or to the danger of the public'.

Before the arrival of the motor-car, the country roads of England had been quiet. The railways had taken away much of the through traffic, and the horse-drawn farm wagons ambled slowly along without fear of the thundering hooves of the mail coach. Cyclists had appeared in great swarms in the 1890s and had caused some alarm, but they were not nearly so fearful as the dragon-like monsters that C.F.G. Masterman described in his *The Condition of England*, published in 1909: 'Wandering machines, travelling with an incredible rate of speed, scramble and smash and shriek along all the rural ways. You can see them on a Sunday afternoon, piled twenty or thirty deep outside the new popular inns, while their occupants regale themselves within. You can see evidence of their activity in the dust-laden hedges of the south country roads, a grey mud colour, with no evidence of green; in the ruined cottage gardens of the south country villages.'

No doubt Marconi covered a few cottage gardens in dust on his trips from London to Poole, though there is no evidence from this era in his motoring career that he got into any serious trouble. For him the motor-car was not only convenient, providing him on land with the kind of freedom he felt on ocean liners: it confirmed his position as part of the upper-class English society in which he now mixed freely, attending *soirées* in London and enjoying invitations from the aristocracy and the wealthy on the south coast.

On one of Marconi's visits to Brownsea Island that summer he was greeted at the pier by a young girl who was gauche and shy in the presence of such a famous man. The ill-kempt nineteen-year-old Beatrice O'Brien, known as 'Bea', was a friend of the van Raalte girls. Marconi, then thirty years old, fell in love at first sight. Bea was the daughter of a most distinguished Irish aristocrat Lord Inchiquin, making her pedigree similar to Marconi's mother's, though more prestigious. The Inchiquins could trace their aristocratic roots back several hundred years, and owned a family seat called Dromoland in County Clare,* as well as a mansion in London. At this time, however, their wealth was draining away.

---

* Dromoland Castle survives as an expensive American-owned hotel.

Beatrice's father, the fourteenth Baron Inchiquin, was in many ways a throwback to the early Victorian period. For Queen Victoria's Jubilee in 1897 he had his own state coach, and at Dromoland Castle he employed a huge range of servants, including one footman to carry the gravy at table, and another to pass the bread sauce. Bea was one of fourteen children, eight girls and six boys. All the boys were sent to public school and then to Oxford or Cambridge, in the hope that they would gain some professional expertise, as only the eldest son would inherit the estate. The girls were given very little education, as the ambition for them was simply that they should marry well. As a child Beatrice mixed with the most eligible of young men. Her mother took her and her sisters on a giddy social round of the finest houses in England, and a favourite summer stay was at Holkham Hall in Norfolk. Close by was Sandringham, the country estate bought by Edward VII when he was Prince of Wales, and Beatrice was invited there to play with young Prince George, who would become King on Edward VII's death in 1910.

When Beatrice met Marconi at the pier on Brownsea Island she already knew something about him. Her father had told her and her sister Lilah about the remarkable young Italian inventor when he was first making a name for himself on the Isle of Wight and in his coverage of the Cowes and Kingston regattas. Lord Inchiquin had died in 1900, before Marconi had become truly world famous, but Beatrice recalled her father's admiration for the inventor. Before coming to Brownsea Island Beatrice had been staying in Chirk Castle in Wales with another aristocratic family, the Howard de Waldens. There she had had her first, and rather frightening, experience of motoring, as a passenger when young Howard de Walden ploughed into one of the ancient walls of the castle at fifteen miles per hour.

Marconi visited Brownsea Island regularly while Beatrice was staying there, then bought a ticket to a great ball held at the Albert Hall which he knew she would be attending. He searched for her in the bejewelled throng, and when he found her he proposed. She hesitated, then a few days later invited him to tea in London and turned him down. Though he was a celebrated figure, Marconi was not necessarily regarded as a great catch. Even though Beatrice's mother had eight daughters to marry off, she certainly favoured people of breeding rather than foreigners, however inventive. Josephine Holman might have said she preferred the wireless wizard to 'a King', but that was not the view of Lady Inchiquin.

In order to free himself to propose to Beatrice, Marconi had asked Inez Milholland to agree to break off their engagement. Now he was unattached again.

Marconi continued his gruelling schedule, with experimentation at the Haven Hotel, another stay in New York in the autumn to fight legal battles, and a trip to Italy, where he was rumoured to have become attached to an Italian princess. As far as Beatrice was concerned the newspaper stories about her suitor's new attachment put an end to what for her had been a very difficult time. When she went to stay with the van Raaltes again she asked that they should make sure Marconi stayed away.

Though Marconi continued to work almost continuously, the companies that carried his name had now grown to such an extent that it was quite impossible for him to attend to every aspect of the work. His status and expertise were needed for the fighting of patent suits and in negotiations with governments, and he had less and less time for technical innovations. By December 1904 there were sixty-nine Marconi shore stations in Britain and abroad, and 124 ships had been equipped with wireless. In addition to the British Marconi Company, the International Marine Company supplied ships, and there were subsidiaries in America, Canada and various European countries. Collectively, with several hundred staff making and operating equipment for ship and shore stations, this was probably the most extensive and best-organised wireless telegraphy business in the world at the time, rivalled only by the German Telefunken. The crackling spark transmitters and the trusty magnetic receivers, known affectionately as 'Maggies', which Marconi had first devised in 1902 to replace the coherer were no longer at the forefront of wireless development, but they were easy to turn out in the factory and were, above all, reliable. No other wireless telegraphy system was so well advanced in terms of organisation, so sure to deliver the goods.

Finally in 1904 the British Parliament put the Marconi Company on a solid footing: a new Act, which came into force the following year, gave the Post Office the responsibility of issuing licences for wireless telegraphy, and Marconi was duly awarded them. The Germans were still seething with resentment because Marconi operators would not exchange messages with others using Slaby-Arco equipment on ships. Despite the pressure brought on the Marconi companies to comply with German demands, they refused, though their old argument that communication was technically impossible was not tenable.

The company just kept ahead of the field. For Marconi himself it had been a very tough year. His widowed mother wrote to him on notepaper edged in black, wondering how he was. Often she wrote to one of his employees, Mr Kershaw, to ask for news of her son. Marconi was obsessed with the problems of his transatlantic service and the new Cape Breton station, which locals called 'Marconi Towers'. And he had fallen in love and been rejected. On his return from New York at the end of the year it must have seemed to him that the only pleasure in his life was motoring, though even that was hazardous: the loyal Kemp noted in his diary a series of punctures.

Then, in December 1904, Kemp mentions the Haven Hotel station being visited by a Mr and Mrs O'Brien, Bea's brother and mother. After Christmas there is a report that Mr Marconi has left for London in his car – with Miss O'Brien. The inventor and the anxious aristocratic teenager were speeding along in a white Mercedes, churning up mud on country roads, frightening horses, defying convention and heading for a future together as magical and uncertain as anything so far in Marconi's astonishingly eventful life.

28

On the American Frontier

At his experimental station at Brant Rock, Massachusetts, Reginald Fessenden took nearly a year to build a transmitter to his own specifications while living with his wife Helen and young son in a remote cottage. His assistants lived in another cottage, and the whole site was fenced off and carefully guarded the year round. They were especially vigilant in the summer, when holiday-makers headed for the coast and were infuriatingly inquisitive about the strange, rocket-like aerial that was under construction. It was possible to climb inside this tubular steel mast to the top, and take in the view of the surrounding countryside and the sea. When the portly Fessenden made the attempt he managed to ease his way to the summit, but found he was jammed when he tried to slither down. To the amusement of his assistants, 'the Old Man' had to take off his clothes and cover himself in axle grease before he could slither back to the ground.

Fessenden was very much a hands-on inventor. But he was also the most scientific and theoretically inclined of the wireless wizards then at work on a commercial system. His ideas about how to transmit and receive electro-magnetic waves were genuinely new. Instead of a spark creating dots and dashes, he believed a very high-powered 'alternator' could be used to send out a continuous wave – which held out the exciting possibility of transmitting speech. What Fessenden really wanted to achieve was transatlantic telephony. His electrolytic receiver, or 'barreter', was much more sensitive than Marconi's magnetic detector. But that proved to be its greatest weakness, for it picked up so much atmospheric static that in certain conditions signals were effectively drowned out. This problem almost drove Fessenden mad, and he was desperate to find a solution.

Whereas Marconi ignored the talents of the eccentric Oliver Heaviside even though he lived only a short motor-car ride from the Haven Hotel, Fessenden did not. He wrote to Heaviside offering to employ him as a technical consultant and enclosing a fee of £100. Heaviside, now considered by the locals to be completely crazy, turned the offer down and refused to accept the money. Fessenden tried again, but Heaviside regarded his work in electro-magnetism to be more or less finished, and refused to apply his inventive mind to Fessenden's problems with atmospheric interference.

Fessenden had been looking for a site on the other side of the Atlantic, for which he needed a licence from the British Postmaster-General. It was no doubt with a wry smile that the political head of the Post Office, having surveyed the western coastline, allotted Fessenden just about the remotest spot possible. West of Glasgow, the Scottish Highlands rise in a jigsaw of sea lochs, glens and rugged bays. The Mull of Kintyre is a long finger of heather-covered land which juts southwards, bounded by the River Clyde estuary on one side and the Atlantic on the other. Fessenden was given a licence to operate from the village of Machrihanish, on the western seaboard of the peninsula, a place with no railway; his men made the last leg of the trip from Glasgow in an open horse and cart. But, bit by bit, the station at Machrihanish was put together, tethered to the rocky ground with huge cables to sustain it against Atlantic storms. At least nobody was going to spy on Fessenden's work in such a remote spot, and there was none of the paranoia that pervaded the station at Brant Rock. Fessenden's millionaire backer Hay Walker Jr had warned him: 'De Forest and other obnoxious persons should be prevented from seeing what you are doing.'

Though de Forest had a reputation for stealing inventions, making slight adjustments to them and calling them his own, he was at this time quite preoccupied, and in reality a threat to neither Fessenden nor Marconi. If anything, it was his own loyal staff who were endangered by de Forest's shoddy approach to wireless telegraphy. After his triumph at the World's Fair in St Louis, the US Navy had set de Forest the task of establishing a series of stations in Pensacola and Key West in Florida, and on a number of West Indian islands, including Cuba. Frank Butler, who had attached himself to de Forest at the World's Fair, was sent out to build the Cuban station, on a bleak promontory composed chiefly of coral. After a gruelling train journey from Havana to Santiago, then a boat trip and a slog through undergrowth, Butler reached the jungle of Guantanamo.* He had seconded to him some US Navy men, including a government inspector who sat observing what was going on from 9 a.m. to 5 p.m. every day without ever uttering a word. Butler hired local labour, including a renegade Frenchman who acted as chef and general dogsbody. They were continuously attacked by insects, scorpions, wild cats and snakes. Not long after the station house was built, the motto 'Abandon hope, all ye who enter here, for verily this *is* hell' was nailed over the door.

Butler kept a diary, which included the following entries for the summer, autumn and winter of 1905:

> June 5th: Big 50 H.P. motor generator blew up, damaging armature.
>
> June 26th: Killed an 8-foot Moha snake in back yard. This was the cause of so many of our chickens disappearing.
>
> July 13th: Terrific storm 2.30 am. Lightning struck station bursting an entire room full of condensers – just finished after two weeks of hard work – throwing oil and plate glass all over the room and into the walls.

* The same Guantanamo to which the United States took Afghan prisoners of war in 2001–02.

August 21st: Small cyclone struck us

August 31st: Lightning struck the station at 4.15 pm blown up one set of condensers.

September 5th: No fresh water. Had to drink salt water all day.

September 24th: Another entire span of 15,000 feet antenna wire blew down.

September 27th: Touched off station again and blower motor blew up.

October 8th: Herd of horses from workmen's camp broke corral in night and demolished the guy wires on the entire aerial spans twisting wires badly.

October 15th: Earthquake at 4.43 pm while eating supper.

November 17th: Heard Key West and Pensacola for the first time.

December 15th: Big two-ton transformer blew up.

When the Guantanamo station did briefly manage to operate, the tropical climate generated such static that it was difficult to read the signals. But Butler kept to his post, and was rewarded with encouraging letters from de Forest. On 9 August he was gratified to receive a note which read: 'You certainly are the star martyr to the wireless cause at present and have our fullest sympathies – if those will do you any appreciable good. None of us are too happy or enjoying flowery beds of ease. It is a tough problem ... but will keep trying new stunts until it is solved. "Never say die" and "You can't stop a Yank" are the two cardinal mottoes of the wireless bunch you know.'

De Forest himself believed that he was about to become rich. He had tasted the kind of life he yearned for at the World's Fair, although this brief opulence proved to be a chimera, just one of Abraham White's stunts to convince potential investors of the success of the business. De Forest had also turned his mind to romance, but in this, as in his choice of technology, he was sadly lacking in judgement. In November 1905 he married a woman called Lucile Sheardown, only to discover that she was the mistress of another man and that she refused to consummate the marriage. Calling her a 'harlot' in his diary, he separated from her after five months.

Very rapidly, de Forest's life fell apart. With Frank Butler struck down by yellow fever in Cuba, and the US Navy contract running into serious trouble because the equipment was so unreliable, de Forest crossed the Atlantic as a second rival to Marconi in the race to establish wireless telegraphy between the United States and Great Britain. The Marconi Company had advance news of these plans, and was prepared to see him off with legal action. But it was not necessary: the few experiments de Forest carried out ended in total failure. By April 1906 he was back in New York, where he was presented with the prospect of a spell in jail. He and Abraham White were judged to be in contempt of court for continuing to use the electrolytic detector patented by Fessenden without payment. De Forest was advised to flee to Canada for a while, but White paid a fine and kept them out of prison. Then he fired de Forest, leaving him with only $500. Bitter at his treatment, de Forest wrote in his diary: 'He [White] has made of me these years an office boy, a traveller about the country to meet people, to talk glowing prospects, to build and operate impossible stations, so that his stock agents might reap large commissions, while he stole the residue.' In desperation de Forest asked Hay Walker if he could join Fessenden's National Electric Signalling Company (NESCO). There was as much chance of that as of him signing up with Marconi, who was, in any case, preoccupied with his personal life.

29

Marconi gets Married

The whole of the front page of London's *Daily Mirror* on 16 March 1905 was devoted to photographs of Guglielmo Marconi and his bride, the Hon. Beatrice O'Brien, who were to be married that day at St George's church, Hanover Square, Mayfair. Although the paper showed a good deal of interest in the details of Beatrice's 'lovely and very uncommon toilette', its headline was 'Inventor of Wireless Marries Today'. Reproduced on the front page was a marconiogram* sent in Morse to the couple from Fleet Street which read: 'TO THE CHEVALIER GUGLIELMO MARCONI HEARTIEST CONGRATULATIONS TO YOURSELF AND BRIDE FROM THE DAILY MIRROR'. Each word was written above the dots and dashes of the Morse message __/.__/.__/.__./ ___/__./.., spelling 'Marconi'.

'Extraordinary interest is being evinced in the event,' wrote the *Daily Mail*, 'for the bridegroom, who has given a daily paper to Atlantic liners and promises annihilation to the cable companies, and, if the popular story be true, has compelled the Chinese to compose a special prayer for protection against him, has friends in every quarter of the globe.' Wedding presents were listed, with an estimate that they must be worth £20,000 – perhaps £1 million at

---

* The wireless telegraph messages sent and received by the operators of Marconi's companies.

today's values. Marconi had given Bea a beautiful sealskin jacket and a magnificent diamond tiara. He showered her with jewellery; but no mention is made in the papers of the bicycle which only close family knew was among Marconi's gifts to his bride.

All the newspapers carried an account of the wedding the following day. Under the heading 'Cosmopolitan Throng at St George's', the *Mail* reported: 'A vast crowd of onlookers was outside; for "the man in the street" has been interesting himself greatly in the wedding of the great hero of wireless telegraphy.'

The church was packed, and there was a great crush of people outside. In Edwardian London 'society weddings' usually attracted a crowd, for the aristocracy still had something of the glamour of film and pop stars today. Mostly women gathered to take a look at the fabulous dresses of the bride and bridesmaids, and the female guests. None of that was lacking from Marconi's wedding, as St George's filled with members of the aristocracy, but the *Daily Mail* reporter thought those straining their necks for a glimpse of the couple were unusual. 'The police outside found themselves obliged to cope with a rather obstreperous crowd, in which, oddly enough, the feminine element for once found itself decidedly in the minority. Tall hats and black coats surged up and down the street in as resolute an attempt to see the arriving wedding guests as any women have ever displayed on similar occasions.'

The society magazine *Vanity Fair* honoured Marconi with a caricature to mark his wedding, and defined him for its fashionable readers.

> The true inventor labours in an attic, lives chiefly upon buns, sells his watch to obtain chemicals, and finally after desperate privations succeeds in making a gigantic fortune for other people. Guglielmo Marconi invented in comfort, retained any small articles of jewellery in his possession, and never starved for more than five hours at a time. Therefore he cannot expect our sympathy as an inventor, though he may excite our wonder as an electrician.

He is a quiet man, with a slow deliberate manner of
speech, and a shape of head which suggests an unusual
brain.

The magazine insisted on calling him 'Bill', and was not so
overawed by his reputation that it could not poke a little fun: 'He is
a hard worker, displaying the greatest resolution before unexpected
difficulties. He rides, cycles and motors. Of music he is a sincere
admirer. Being half an Irishman his lack of humour is prodigious.'

The union of a brilliant inventor and an aristocratic young lady
seemed perfect, but the path to the wedding had not been smooth.
It was Florence van Raalte who mischievously brought the couple
together after Beatrice had turned Marconi down. Charles van
Raalte had assured Bea that Marconi would not be invited to
Brownsea while she was there, but his wife broke the agreement,
and the two found themselves once again together on the romantic
island. Under its influence Bea fell head over heels for Marconi.
When she announced that she wanted to marry him, her elder
brother Lucius, now head of the family, and her mother declared
that he was an unsuitable match. No doubt the van Raaltes and
others who had befriended Marconi spoke up for him. He was
almost a gentleman, and at least half Irish-English, even if he
stubbornly held on to his Italian nationality. The *Daily Mail* was
at pains to point out that in St George's church Marconi spoke
the wedding vows 'without a trace of an Italian accent'.

In the end Lucius, now Lord Inchiquin, gave her away.
Marconi's best man was his elder brother Alfonso, and his mother
Annie mingled proudly with the glittering array of lords and ladies.
Old Giuseppe had died a year too soon to see his son fêted in
London, but King Victor Emmanuel sent a personal letter to
Marconi signed 'yours affectionately', and the Italian Ambassador
was among the wedding guests. One of the newer European off-
shoots of the wireless enterprise, the Belgian Marconi Wireless
Company, sent a large jar with the message that they hoped there
would be 'no family jar' in the Marconi household. The pun was

unfortunate: from the very beginning this would not be an easy marriage.

After a reception in London at which the fabulous wedding gifts were displayed, Marconi and Beatrice travelled to Ireland, where they had been offered two weeks in the Inchiquin family pile, Dromoland. It was now a sad and lonely place, which had seldom been lived in since the death of Beatrice's father five years earlier, and she missed the gaiety she had enjoyed there in her childhood. Still, she remained vivacious, bubbly and naïve – but these very qualities which attracted Marconi also made him absurdly jealous and protective. He became moody, taking long walks on his own.

The newlyweds returned to London before the two weeks of the honeymoon were up so that Marconi could get back to work. They stayed for a time in the fashionable Carlton Hotel,* and naturally enough Bea liked to walk around the West End when her husband was out on company business. But Marconi would not tolerate such small freedoms, imagining that his young bride was taking an interest in other suitors. He could not abide the suspicion that his wife might be unfaithful to him – though later he himself was to have a reputation as a philanderer – and did his best to keep her out of the public eye and under lock and key.

In May they took the liner *Campania* to New York. It was, as always, a working trip for Marconi, who was testing the signals from Poldhu. At first Beatrice enjoyed the social life in first class, but she was soon to get a lecture from her husband about 'flirting' with other passengers. To distract her from romance Marconi taught her Morse code, and she took on the tasks – formerly attended to by his mother – of tidying his clothes and darning his socks. In New York the couple were entertained at a round of parties, and had lunch with President Theodore Roosevelt before

---

* The Carlton Hotel, in the Haymarket, was opened in 1899. It was run initially by César Ritz, with the celebrated Escoffier as the chef. Ritz reputedly suffered a nervous breakdown when Edward VII's coronation was postponed, as he had prepared all the food for the banquets. The Carlton was demolished in 1957–58.

travelling to the bleak landscape of Cape Breton, where Marconi continued to grapple with the problem of getting his transatlantic service working. Beatrice had to share the station house with Richard Vyvyan's wife Jane, and the two found life together difficult to begin with. There was no social life for Beatrice here, though as a bachelor Marconi had been rumoured by the newspapers to have had many woman friends on Cape Breton.

When Marconi returned to England on the *Campania* for further tests of his transmitters he left Beatrice behind for three whole months. Neither of them knew when he left that she was pregnant, and she was sick much of the time, which made her life on Cape Breton even more miserable. When she came back to England Marconi at first put her up in the Poldhu Hotel, a lonely and dull place in winter. Concerned for her daughter's well-being, and in the belief that Marconi could afford it, Lady Inchiquin found her a rented house just off Berkeley Square in London's Mayfair.

Although he accepted this, Marconi was actually in deep financial trouble, living a life he could not afford. To keep the company going he had put most of his own money back into it. If he had fallen in with the kind of financial backers Fessenden and de Forest relied on, it is doubtful he would have survived this crucial time. But his board of directors remained loyal to him as he asked for more and more money to sustain his married life and to invest in larger and more powerful transmitters.

Beatrice gave birth to a daughter in February 1906. The child was named Lucia, but before she could be baptised she died of an infection. Marconi wrote to his mother: 'Our darling little baby was taken away from us suddenly on Friday morning. (I was at Poldhu at the time and only got here when it was all over). She had been very well all the time before and the doctor said she was a more than usually healthy baby. On Thursday night she had what was thought to be a slight attack of indigestion and at about 8 a.m. on Friday morning an attack of convulsions and all was over in a few minutes. Bea got a most awful shock and she is now very weak . . .' Because the little girl was not baptised, Marconi had to use all his influence to get her a Christian burial. Finally there was a brief service at a cemetery in west London.

With Beatrice sick and grieving, and the pressures of work intense, Marconi himself fell ill. He had suffered before from feverish attacks, which were believed to be a recurrence of a bout of malaria. Delirious for much of the time, he lay in bed in the rented house in London, fearful of any medicines he was given, always checking the labels, and rejecting many because he thought they contained something which might harm him. With black humour he cut out undertakers' advertisements and propped them on a table at his bedside. He railed against English doctors and nurses, who he said treated him like an idiot, and insisted on being attended by a Dr Tallarico from the Italian Hospital in London.

To be struck down by illness at such a time was a torment for Marconi. After persistent experimentation with the shape of the aerial at Glace Bay there was now the prospect of a breakthrough in his attempts to make the transatlantic link work reliably. He and his engineers had discovered that if a pattern of very long wires was stretched out horizontally above the ground, facing in the direction signals were transmitted and received, performance improved dramatically in both daylight and night-time hours. The so-called 'directional aerial' was huge, and demanded acres of land. There was room at the new 'Marconi Towers' site at Cape Breton, but the clifftop Poldhu station, which had looked so imperiously grandiose two years earlier, could not be extended. Reluctantly, the company had agreed that a larger site had to be found, and to defray the cost they had asked the British Admiralty if it would contribute. When help was refused, they went ahead anyway, and to increase the chances of success decided on a site nearer to Cape Breton than Cornwall. Land was acquired at Clifden, on the west coast of Ireland, and a station and directional aerial constructed which dwarfed the men who worked on it. The storage batteries for the power plant were the height of a four- or five-storey building.

Marconi was becoming aware that he was reaching the limits of his spark technology. In Cape Breton the flashes and thunderclaps

of the new station attracted so much popular attention that the
local railway company made a special stop close by, so that passengers could watch the tongues of bluish-white flame which flashed
when the Morse key was pressed, and, with hands over their ears,
feel the thunderclap and the crackling of the aerial wires. Marconi's
equipment had not so much 'snatched the thunderbolt' as emulated
the thunderstorms which inspired the experiments of his boyhood
hero Benjamin Franklin. Yet no other inventor had devised a
system of wireless telegraphy to replace Marconi's, whatever claims
they might make. The spark transmitter and the coherer and
Maggie detectors had not yet played out their historic role. As
Marconi lay on his sickbed the first real test of the value of wireless
in wartime was unfolding dramatically in the Far East, as the
Russians made a last-ditch attempt to defeat the Japanese.

30

❧❧❧❧❧

Wireless at War

Long after Captain Lionel James had been expelled by the Japanese, the Russo–Japanese war had rumbled on with land and sea engagements, but with no decisive action by either side. It was in desperation that Tsar Nicholas and his military chiefs in St Petersburg took the decision to despatch fifty-nine ships from the Baltic port of Libau to join the remnants of the Far East fleet in Vladivostok. To reach the safety of that naval base they had to steam through the Straits of Tsushima, which divide southern Korea and Japan, where they knew the enemy would be waiting to ambush them. In the spring of 1905 the entire Russian Baltic fleet was nearing the East China Sea after an eighteen-thousand-mile voyage which had begun in mid-October the previous year. The Russian navy, stationed at Port Arthur and Chemulpo on the Yellow Sea and at Vladivostok, had had the worst of the frequent naval encounters, and had lost a large part of its fleet either sunk or captured.

The navies of all the major powers awaited the outcome of this confrontation with particular fascination. For the first time in history both protagonists were equipped with wireless, and nobody knew how this would affect the fighting, or what strategies would be adopted.

In command of the Russian fleet was Admiral Zinoviy Petrovitch Rozhestvensky, who had achieved rapid promotion. When Tsar

Nicholas and Kaiser Wilhelm had staged their impromptu meeting in 1902 while awaiting Edward VII's postponed coronation they had been entertained by displays of naval prowess. Rozhestvensky was the young officer in charge of a battleship which gave a demonstration to the two imperial leaders of firing at moving targets. After observing three hours of faultless and unruffled command, Kaiser Wilhelm said to his despised cousin the Tsar: 'I would be glad to have in my navy officers as efficient as your Rozhestvensky.'

The Russian army had bought some Marconi equipment, which was set up in the harsh conditions of Siberia then nearly destroyed by Orthodox priests who insisted on anointing it with holy water. But the Russian navy had been more patriotic, adopting the Popov-Ducretet system* for its ships. Admiral Rozhestvensky had had some experience of this equipment, and was unimpressed. In theory the Russian fleet had the key secret weapon in the forthcoming conflict, a wireless station established on the cruiser *Ural* with a range of seven hundred miles. The equipment, which had been bought from Germany, would enable the Baltic fleet to contact Vladivostok as it approached the danger zone of the Straits of Tsushima, and alert the surviving ships there to attack the Japanese from the opposite direction.

Nobody had sold the Japanese any wireless equipment. The general belief was that the system they had developed was a straight copy of that demonstrated by Marconi at La Spezia in 1897, and that it had been improved very little. It was certainly way behind Marconi's own system at the time. But it worked efficiently over a range of up to six miles, and with shore stations set up along the Korean coast, on anchored battleships and the western shores of Japan, a large area could be covered by a relay system from one station to another. The Japanese had the Straits of Tsushima covered, and waited and watched for the Russian fleet, their wireless antennae ever alert for the crackle of the enemy's Morse spark.

* Devised by the Russian Alexander Popov (also spelled Popoff) and the French inventor Eugène Ducretet.

Which of three routes, winding between islands in the narrow seas, would Admiral Rozhestvensky go for? How would his fleet's wireless be used to outmanoeuvre and engage the enemy? To the alarm and disappointment of many of his own officers, Rozhestvensky decided on total wireless silence from his ships, reasoning that any signals they sent would instantly give the Baltic fleet's position away to the waiting Japanese. His aim was not to engage the Japanese but to make it through the Straits to Vladivostok unnoticed. All the Russian wireless operators were asked to do was to listen for Japanese signals.

Rozhestvensky chose to go through the Straits of Korea, and his ships approached them in two lines on 25 May 1905. The weather was rough, with poor visibility. The Russian fleet soon picked up wireless signals from Japanese ships, which became stronger as they headed north-east towards Vladivostok. On a misty morning three days later a Russian hospital ship sighted another vessel and signalled to it, unaware that it was a Japanese cruiser, the *Shinamo Maru*, which had been tailing it. As the mist cleared and the *Shinamo Maru* saw the great line of the Baltic fleet it began to send warning messages, but they did not reach any Japanese stations, which were too far away. However, the Russian fleet picked them up instantly. Guns were trained on the *Shinamo Maru*, but Admiral Rozhestvensky gave no order to fire, possibly fearing that to do so would give their position away.

The captain of the *Ural* signalled in semaphore for permission to jam the Japanese cruiser's wireless using his powerful equipment. This request was turned down, and the Russian fleet ploughed on at a steady nine knots towards Vladivostok, taking a middle course through the Straits of Korea. It was then that the *Shinamo Maru* came close enough to the rest of the Japanese fleet to report the exact position and route of the Russian ships. Rozhestvensky, knowing that his intended passage must have been reported to the Japanese commander Admiral Togo, could have changed course. Many urged him to do so. But instead he steamed straight into Togo's well-prepared ambush. The first volley of Japanese shells killed all Rozhestvensky's officers, and left him unconscious and badly wounded. Only three of the fifty-nine Russian ships got through to Vladivostok: the rest were sunk or captured, and Rozhestvensky himself was taken prisoner. The confident young officer admired by the Tsar and Kaiser Wilhelm had misunderstood the power of wireless in warfare, with disastrous consequences.

In his defence, it was said that Rozhestvensky's experience of his inferior wireless equipment made his decision not to use it justifiable, and that the outcome of the conflict, given the strategic advantage of the Japanese, might not have been very different had he made use of wireless, as many of his officers had wanted him to. As for the Japanese, they had adopted early on an unsophisticated system which worked efficiently within its limitations, and used it to the full. It was some consolation for the Marconi Company that, having been beaten to the reporting of war by wireless, the war itself was won with equipment copied from their own design. And from that time on all the navies of the world recognised that they had to come to terms with wireless, and that what they needed above all was a system they could have faith in.

By 1904 the United States government had set up various committees to consider the uses of wireless, but the US Navy continued to dither over which was the best system to invest in. Naval shore stations had been set up the length of the eastern seaboard, and trials went on more or less continuously in an effort to overcome the problem of interference between transmitters whose wavelengths overlapped. Then, in 1905, the navy began to experience a problem it had not anticipated. One of the beauties of Marconi's invention was that anybody with a few dollars and a bit of ingenuity could get in on the act, and amateur wireless enthusiasts had set up makeshift stations of their own in attics and bedrooms all along America's east coast. Some of them had equipment more powerful and efficient than the navy itself. And there was nothing the American amateur enjoyed more than hacking into the wavelengths of the US Navy.

31

❧❧❧❦❦❦

America's Whispering Gallery

A new shop, the very first of its kind, opened in 1905 at 233 Fulton Street, New York. It was called the Electro Importing Company, and was run by a recent immigrant from Europe, Hugo Gernsback, who had arrived in the United States the previous year with a design for a new dry battery which he hoped would make him a fortune in the New World. Gernsback had been brought up and educated in Luxembourg, studying at the École Industrielle and later at university in Bingen, Germany. Born in 1884, he was ten years younger than Marconi, and had a very different vision of how wireless telegraphy might develop. Gernsback was not concerned about competing with cable companies or sending signals over thousands of miles; he foresaw that cheap, easy-to-build wireless equipment might have a popular appeal among boys and young men like himself. At the end of 1905 he bought space in the magazine *Scientific American* to advertise what was almost certainly the first do-it-yourself wireless station kit. He called it the 'Telimco Wireless Telegraph Outfit'.

The Telimco kit had its own spark transmitter and coherer-receiver, and was guaranteed to work up to a range of one mile. It was in fact very like one of Marconi's boyhood magic boxes, and in no way represented any kind of technological breakthrough. But Gernsback, who also sold do-it-yourself Telimco X-ray kits, had recognised a fact that Marconi had failed to appreciate despite his own background as a schoolboy inventor: homespun wireless telegraphy could be fantastic fun. From the attics and playrooms of suburban Boston or New York the amateur wireless enthusiast could tune in to the exciting clandestine world of the airwaves, which buzzed every night with hundreds of messages. He – there were very few girl amateurs – could learn the secret language of Morse code and use it to send messages to others with home-made wireless stations. Gernsback got in at the beginning of a craze which was to sweep the east coast of North America, and had a profound effect on the way wireless was to develop.

In Great Britain, amateurs had to take out experimental licences from the Postmaster-General after the Wireless Act became law in 1905. But in the United States there were no regulations, and the amateurs were to enjoy a brief period of total freedom, which they exploited with great enthusiasm, ingenuity and a considerable amount of irresponsibility. They had no need to worry about legal action, as they were small fry; if they infringed Marconi's tuning or other patents, there was no point in a big company pursuing them.

Very rapidly the amateurs learned how to make much more powerful sets than Gernsback's Telimco starter kits, pillaging all kinds of equipment, much of it designed for quite other purposes. They used old photography plates wrapped in metal foil to make batteries, generated sparks between old brass bedstead knobs, and found that the electrical ignition coils from Model-T Fords pro-vided transmitter power. Cannibalised electric fans could be used to generate a string of sparks. Crude loudspeakers were made with rolled-up newspapers, and when Quaker Oats brought out a cylin-drical cardboard package it made an ideal drum around which the wires for a crude tuning system could be wound. Umbrellas' ribs could be used for aerials, and the power of transmitters could be bumped up by an illegal tap into street power cables. The most difficult and expensive item to acquire was a headset: these were often stolen from public telephone boxes.

Very soon the ingenuity of the amateurs created a vibrant

community of backroom stations on the eastern seaboard which represented the greatest concentration of wireless telegraphy activity in the world. This began to capture the interest and imagination of the popular newspapers and magazines, who were thrilled to discover that youthful enterprise had taken hold of a novel technology and fashioned an entirely new and exciting hobby. In November 1907 the *New York Times Magazine* ran a full-page lead story headlined 'New Wonders with Wireless – and by a Boy!' The 'boy', twenty-six-year-old William J. Willenborg, had built himself a wireless station in his parents' home in Hoboken, New Jersey. Photographed in jacket and tie and with an expression of cool determination, he was honoured with a large oval portrait and surrounded by pictures of his apparatus and rooftop aerial. The reporter who had been invited to join him for an evening's eavesdropping was ecstatic with the excitement of it all: 'For intrigue, plot and counterplot, in business or in love or science, take to the air and tread its paths, sounding your way for the footfall of your friend's or enemy's message. There is romance, a comedy, and a tragedy yet to be written.'

Willenborg demonstrated the command he had of the airwaves. He showed the *Times* reporter how he could jam the signals coming from the Atlantic Highlands station that Marconi had originally had built to take in the 1899 America's Cup results. In frustration the professional operator tapped out: 'Lay off New York!' When Willenborg impishly jammed him again, the operator Morsed: 'Go to hell.' When Willenborg leaned on his key again to jam Atlantic Highlands the operator finally gave up, much to the amusement of the 'boy' amateur and the reporter.

Not all the amateurs were able to transmit signals and sabotage US Navy and Marconi transmissions. Many were content just to listen in, and for them the great breakthrough was the marketing from 1906 of 'crystal set' receivers. The discovery that certain minerals acted as 'semi-conductors' of electro-magnetic waves had been made around the turn of the century by the German academic Professor Ferdinand Braun, but nothing practical came of it. Then simultaneously two men working in the American wireless industry patented the 'crystal set', in which a thin wire or 'cat's whisker' was brought into contact with the surface of a piece of carborundum or silicon to act as a receiver. With a crystal set, a bit of tuning and a telephone set, the amateur could tap into just about any wireless signal.

When, after years of frustration and toil, Marconi finally had his transatlantic telegraphy system between Cape Breton and Clifden working reliably in October 1907, there was no triumphant public announcement. It was primarily a service for newspapers, and the company policy was simply to get it running and to hope there would be enough custom to make it pay. The messages transmitted from Cape Breton and Cape Cod provided the American amateurs with great interest, as they could easily tap into them and get a thrilling sense that they were connected with the very heart of an exciting world of urgent news. The night sky had become one great 'whispering gallery', as an article in *Scientific American* put it, every word of it tapped out in the romantic dots and dashes of Morse code. But it would not be long before there was music and speech as well on the airwaves.

32

A Voice on the Air

In the late summer of 1906, at New York's fashionable Café Martin on 26th Street between Fifth Avenue and Broadway, diners were entertained to a performance of popular pieces by classical composers. There were no musicians to be seen, nor was there an Edison phonograph on show: the renditions of Rossini, Chopin, Grieg and Bach emerged from foghorn-like speakers arranged in the rooms. The two pianists playing the music were some way off, in a building opposite the Metropolitan Opera House on Broadway, level with 39th Street. The notes the café's customers heard were generated electronically by a huge bank of alternators, each of which had been set to hum at a different pitch, and travelled down telephone wires to be 'broadcast' from the speakers in the Café Martin.

The machine which produced this synthesised, piped music was called a Telharmonium, and weighed two hundred tons. It made a deafening hum, and was housed in the basement of the music hall in which the pianists were playing. The Telharmonium, or Dynamaphone as he sometimes called it, was the brainchild of a Washington DC lawyer named Thaddeus Cahill. An earlier version of this giant electronic music generator, which Cahill described as a synthesiser, had been patented in 1898. Cahill's prototype weighed a mere seven tons, and worked well enough for him to get financial backers to form a company called the New England Electronic Music Company.

With his new funds Cahill refined and enlarged his machine. In search of audiences, the company moved to New York, and one of Cahill's backers, Oscar T. Crosby, formed the New York Electronic Music Company. The giant machine with its 145 generators or alternators was dismantled and loaded onto thirty-two railroad cars to be transported to New York, where it was carried in horse-drawn wagons to Broadway. A deal with the New York Telephone Company allowed the laying of lines to cafés and restaurants to provide the clientele with a kind of early Muzak.

At first the novelty of the Telharmonium performances attracted widespread interest and admiration, but it was soon discovered that the powerful signals jumped from the company's lines to those of ordinary telephone subscribers, who were furious when disjointed bursts of Handel or Bach interrupted their conversations. However, Cahill's invention survived just long enough to attract the attention of Lee de Forest, who was experimenting with a new kind of receiver which could receive speech and music as well as Morse code.

All those concerned with wireless around the world were searching for more efficient ways of transmitting and receiving electromagnetic signals. Among them was Ambrose Fleming, whose laboratory at University College, London was littered with rejected experimental aerials, batteries and generators. As he racked his brains for an idea for a better detector than the 'Maggie', he recalled that when he worked for Thomas Edison an intriguing observation had been made about the flow of electric particles through a vacuum lightbulb. It was known as the 'Edison effect', and in essence was the fact that the current between the filament and the electrode within a bulb would run one way, and not the other. In that sense the bulb acted like a valve in a water pipe, which closed to the flow in one direction, and opened if the flow was reversed. With a little adaptation a lightbulb could be made to pick up the alternating impulses of a wireless signal and convert it into a direct current which would activate a telephone receiver. Fleming had a dozen valves made to his own design by the Ediswan company, and when they proved effective he ran off along Gower Street to the Patent Office in his green galoshes. That was in 1904, and over the next three years the 'diode' valve was tested out at Marconi stations.

In America Lee de Forest, still seeking his fortune, adapted Fleming's valve to produce a receiver which he called an 'audion'.* With this he made an attempt at transmitting Cahill's Telharmonium music by wireless. But the audion was a very imperfect piece of equipment: the sound quality was dismal, and the performance was hardly pleasing to the US Navy, which found that its cruisers and battleships were picking up ear-splitting renditions of Rossini instead of vital information about manoeuvres. The public too appeared unimpressed by the notion of broadcasting which de Forest was just beginning to imagine. In December 1906 the *New York Times* carried an editorial with the headline 'A Triumph but Still a Terror' in response to de Forest's claims that he was about to make a big breakthrough in wireless telephony: 'There is something almost terrifying in the news … that attempts at telephoning without wires have already attained such success that scientists announce the approach of the time when man will be able to speak without any conducting wire to a friend in any part of the world.'

Out on the snowy wastes of Brant Rock, Reginald Fessenden was still pursuing his ambition to be the first in the field of 'wireless telephony' by transmitting speech. He had asked General Electric to build him high-speed generators which could send out a continuous stream of electro-magnetic signals, and had used these in his attempts to establish a telegraph link between Massachusetts and Machrihanish on the west coast of Scotland, very nearly beating Marconi in the race to cross the Atlantic with intelligible messages. But in January 1906 the quality of reception varied so much that he was often in despair. He wrote to a friend: 'Sometimes the signals are very loud, so that we can hear Machrihanish with the

---

* De Forest added a third electrode, or grid, to Fleming's simple valve. This had the effect of making it possible to amplify high-frequency signals, and led to the development of radio.

telephones six inches away from the ear, but two or three times every month we can hardly hear them at all, which of course is not commercial.' He was plagued too with the same problem Marconi had, of the loss of reception during daylight hours.

Nevertheless, by the autumn of 1906 Fessenden's NESCO remained a serious rival for the Marconi Company, and his excited backers were awaiting the announcement that the transatlantic service was reliable enough to be offered to businesses and newspapers. They promised to ride around Pittsburgh in a cab with their legs hanging out when the news came through. Instead, on 5 December the Machrihanish mast collapsed in high winds. It was a fatal blow to NESCO, which had promised so much. In the aftermath of the disaster Fessenden continued to indulge his passion for telephony. On one occasion during the ill-fated transatlantic tests an engineer in Scotland had been sure he had heard a voice at Brant Rock. Encouraged by this, Fessenden began sending short spoken messages to the crews of New England fishing fleets.

The boats of the United Fruit Company plying between South America and Boston all had either de Forest receivers or those of Fessenden's company, and the US Navy's ships off the coast of Brant Rock were now mostly equipped with detectors of one kind or another. Early in December 1906 Fessenden sent a telegraph message to all these vessels alerting them to two events for which they were to listen out on Christmas Eve and New Year's Eve. These would be the world's first-ever wireless programmes, and would include speech, music and singing. Fessenden then wrote to phonograph companies asking them to donate records by famous performers, and prepared his wife and his assistants to gather with him around the microphone.

Sure enough, on Christmas Eve the wireless telegraphy operators aboard United Fruit ships as they steamed to and from Boston with their cargoes of bananas were startled to pick up not the familiar dots and dashes of Morse code, but the sonorous voice of Reginald Fessenden himself. He was the sole live performer, as his wife and assistants got stage fright at the last minute. Of that historic night Fessenden wrote:

> The program on Christmas Eve was as follows: first, a short speech by me saying what we were going to do, then some phonography music – the music on the phonograph being Handel's 'Largo'. Then came a violin solo by me, being a composition of Gounod called 'Oh Holy Night' and ending with the words 'Adore and be still' of which I sang one verse, in addition to playing on the violin, though the singing of course was not very good. Then came the Bible text, 'Glory to God in the highest and on earth peace to all men of good will.' And finally we wound up by wishing them a Merry Christmas and then saying that we proposed to broadcast again New Year's Eve.

Fessenden was heard as far south as Norfolk, Virginia. The programme went out again on 31 December, with a different phonograph record and one additional voice, that of a brave assistant who had been persuaded to sing a little.

This first 'radio' broadcast in history was heard by only a few fishermen off the New England coast, a handful of naval officers and the very first radio hams. Fessenden continued with the occasional broadcast, and recorded reception at distances of over two hundred miles. But his wealthy Pittsburgh backers Hay Walker Jr and Thomas Given had set their sights on transatlantic telegraphy, and were not interested; they decided to sell out. Even Fessenden himself thought of wireless telephony only as a replacement for Morse code as a means of communication, and surprisingly never thought of broadcasting as a form of entertainment. His pioneer programmes received practically no publicity, and were soon forgotten. The concept of 'broadcasting' took a long time to dawn on the new wireless industry, which is surprising, because it had in effect been invented before Marconi arrived in London in 1896, and was well established in one European city by the early 1900s.

33
❧❧❧❧

The Bells of Budapest

Visitors to the city of Budapest in the Austro-Hungarian Empire in the first years of the twentieth century might hear, as they arrived by paddle-steamer on the River Danube, the noonday hour marked by the firing of a cannon from the Citadella. Budapest was one of the most vibrant cities in Europe at that time, with a bustling street life, smart hotels – one of which was called the New York – and around six hundred coffee bars. Formed by the union of two cities that lay on opposite banks of the Danube, Buda and Pest, in 1873, Budapest enjoyed a great cultural and economic revival in the late nineteenth century, and was considered by many to be livelier than Vienna. One thing the city lacked, however, was accurate public clocks. The midday cannon-shot filled the gap for a population which had reached around 800,000 by 1900.

There was, however, an alternative to the cannon for an accurate time check for an elite group of around six thousand Magyar-speaking citizens of Budapest, and those who moved quickly enough in some of the most fashionable hotels and cafés. All they had to do was pick up a telephone and tune in to a station called Telefon Hirmondo, which announced the exact astronomical time at noon as part of its daily schedule of programmes. In the Magyar tongue '*hirmondo*' was the term for a town crier, and the city's unique telephone broadcasting system fulfilled the function of the old liveried bellower of the latest news, and a good deal more.

Reporting on this novelty in 1901, Thomas S. Denison wrote in the American magazine *World's Work*:

> For a quarter of a century one of the favorite dreams of the modern prophets has pictured the home equipped with apparatus by means of which one can hear concerts or listen to the latest news, while sitting comfortably by his own fireside. This dream is a fact today in Budapest. Music, telegraphic news 'hot' from the wires, literary criticism, stock quotations, reports of the Reichsrath [the parliament] – the whole flood of matter that fills the columns of our newspapers may be had for the mere lifting of a telephone receiver.

Most of the thousands of foreign visitors to Budapest had never heard of Telefon Hirmondo, and could be startled by the abrupt upheavals it caused in public places. In his *Hungary and the Hungarians* (1908) the English writer W.B. Foster Bovill warned: 'You may be seated as I was in the reading-room of one of the hotels or in a large coffee-house, when suddenly a rush is made for a telephone-looking instrument which hangs from the wall.'

Telefon Hirmondo was the brainchild of Tivador Puskas, a Hungarian who worked for a time at Thomas Edison's experimental hothouse Menlo Park in New Jersey. Edison credited Puskas with the invention of a form of telephone switchboard, and clearly had a high opinion of him. As early as 1881 Puskas had delighted visitors to the Paris Exposition Internationale d'Électricité Téléphonique with a 'Théâtrephone' demonstration, in which a concert was transmitted from a hall by telephone. In 1882 he repeated the show in his native Budapest, with a performance from the national theatre to a nearby hall. When Tivador's brother Ferenc acquired the first telephone concessions in Hungary in 1881 he hired the young Nikola Tesla to work on the equipment's design, setting that eccentric inventor on the path to his career in Paris and the United States. In Budapest, Tivador devised Telefon Hirmondo, creating a newsroom and studio to broadcast regular bulletins to a thousand subscribers who rented lines connected to the central telephone exchange. The first programmes went out in 1893, but Tivador died within a few weeks of his triumph. There were teething troubles, and new lines had to be laid, but the service became well established, and lasted for a quarter of a century.

Very little was known about Telefon Hirmondo outside Hungary until articles began to appear in European and American journals in the early 1900s. In June 1907 *Scientific American* carried a vivid account of the operation of the station. One of the big problems with the existing technology was the amplification of sounds to enable them to travel through long lengths of cable. Although the Hungarians appear to have devised some ingenious solutions, the qualities of the old town crier were still called upon. The printed news items were bellowed out in relay by eight readers, who were observed in action by the *Scientific American* correspondent:

> From eight in the morning till ten at night eight loud-voiced 'stentors' with clear vibrating voices literally preach the editor-in-chief's 'copy' between a pair of monstrous microphones, whose huge receivers are facing each other. The news is of all kinds – telegrams from foreign countries; theatrical critiques; parliamentary and exchange reports; political speeches; police and law court proceedings; the state of the city markets; excerpts from the local and Viennese press; weather forecasts – and advertisements . . . In the event of some ultra-important item coming to hand suddenly – a disaster of international moment, an outbreak of war, or the like – it is instantly shouted into the microphones by the stentors . . . So loudly do they shout the news of the world, that a 'solo' of ten minutes quite exhausts the strongest.

For broadcasts of musical performances there were enormous microphones, four feet in diameter, to receive the maximum volume.

Not everyone in Budapest could tune in to Telefon Hirmondo:

it was a service for an elite. Nevertheless, it embodied a fully developed idea of what later became known as broadcasting. *Scientific American* reproduced a full day's programme:

A.M.

9.00 Exact astronomical time.

9.30–10.00 Reading of programme of Vienna and foreign news and of chief contents of the official press.

10.00–10.30 Local exchange quotations.

10.30–11.00 Chief contents of local daily press.

11.00–11.15 General news and finance.

11.15–11.30 Local, theatrical, and sporting news.

11.30–11.45 Vienna exchange news.

11.45–12.00 Parliamentary, provincial, and foreign news.

12.00 noon Exact astronomical time.

P.M.

12.00–12.30 Latest general news, local news, parliamentary, court, political, and military.

12.30–1.00 Midday exchange quotations.

1.00–2.00 Repetition of the half-day's most interesting news.

2.00–2.30 Foreign telegrams and latest general news.

2.30–3.00 Parliamentary and local news.

3.00–3.15 Latest exchange reports.

3.15–4.00 Weather, parliamentary, legal, theatrical, fashion, and sporting news.

4.00–4.30 Latest exchange reports and general news.

4.30–6.30 Regimental bands.

7.00–8.15 Opera.

8.15 (or after the first act of the opera) Exchange news from New York, Frankfurt, Paris, Berlin, London, and other business centers.

8.30–9.30 Opera.

Telefon Hirmondo made money, as the subscription its wealthy customers were prepared to pay easily covered the cost of delivering the service. The broadcast of the exact time was greatly appreciated by watchmakers and jewellers, who were later to benefit from the same service provided by wireless.

Although similar broadcasts were attempted in France, England and the United States, they were mostly mere novelties. In 1894 the Chicago Telephone Company set up a system to read local election results to subscribers. A team of 150 operators were given twelve to twenty numbers each to phone, with the assumption that the subscribers would listen in silence as the results were read out to all of them simultaneously, then replace the receiver with a polite 'Thank you.' But it did not work out like that. Groups listening to each operator's recital of the results did not remain passive, but began to squabble among themselves. The company's general manager, A.S. Hibberd, gave the *Electrical Review* an account of the exciting experimental evening:

> One lady, on being put on the bulletin circuit, directed the operator that if the returns were favorable to the Democrats she was to be immediately cut off, but that if they were Republican she would remain at the telephone all night if necessary. It is needless to say she remained taking the bulletins until the final number (100) had been read to her.
>
> On another circuit one or two Republicans were inclined to hurrah as the bulletins were read, when a number of irate Democrats shouted out that those interfering Republicans must be cut off the wire. On another a lady interrupted very frequently with the statement that she did not care how New York had gone, she wanted to know right off whether Mrs. Jones was elected here in Chicago, and it took the continued efforts of an unknown man with a bass voice to hold her down so that the others might hear the bulletins.

Reactions to this early 'phone-in' reflect a general fear about broadcasting as a concept: that it could be troublesome and intrusive. The most aggravating experiments in the United States were with 'telephone advertising', a bright idea sales people had once a large number of subscribers were linked to any local exchange. In 1909 a woman complained to her local paper in Rochester, New York:

> My telephone is far more of a nuisance to me than it is a convenience, and I think I will have it removed, if I am called up as much in the future as I have been during the past week by theater agents, and business firms, who abuse the telephone privilege, using it as a means of advertising. My hands were busy moulding bread yesterday morning, when I heard the bell ring, and upon responding was told by a woman just gone into business in a Main street building, that she had a fine line of curtains, and other hangings, which she would like me to see. Shortly afterwards an employee of a firm making extracts, solicited my patronage in the same way, and though I told him that I did not wish to be annoyed again, by being called to the telephone to hear of the extracts, the afternoon brought another call from the same firm. Last week a number of my friends and I heard over the telephone of a Shakespearian actor who was to fill a long engagement here, and we were asked by an attaché of the theater to please get our seats early, as there would undoubtedly be a rush for tickets. These are samples of a telephone annoyance that I would like to be freed from.

Despite the success of Telefon Hirmondo in Hungary, the telephone was not the ideal instrument for broadcasting. Wireless was much better, as it would be possible for an audience to tune in and out when they wanted. But, of course, no members of the general public, apart from a few amateur enthusiasts, had a wireless receiver, and there was therefore no audience. The last thing Marconi wanted for his transmissions was 'eavesdroppers'. He was always at pains to point out that once he had solved the problem of tuning

Morse, messages by wireless would be every bit as private as those sent by cable.

Although Marconi's huge transmitters at Glace Bay, Nova Scotia and Clifden in Ireland were reliable enough to provide newspapers with a transatlantic service as an alternative to cable, this alone was not going to make his fortune. He still spent most of his time travelling, and had taken Beatrice with him when he went to Nova Scotia for the inauguration of the Glace Bay transmissions in October 1907. While Marconi continued to enjoy the acclaim of the Canadians, Beatrice was not allowed to socialise much, or even to go fishing or hunting with the ebullient Richard Vyvyan.

The young couple, saddened by the loss of their first child, still had no permanent home and no settled life. When Beatrice became pregnant again, Marconi finally rented a country home, Sunbourne in Hampshire, owned by Sir Richard Harvey Bathurst. They moved into Sunbourne in the summer of 1908, but Marconi spent most of his time travelling between Poldhu, the Haven Hotel and Clifden. If he was working in London, he and Bea stayed at the Ritz. When Marconi returned to North America he left Beatrice in a house he rented in the West End of London. He was in Canada when their second child was born on 11 September 1908, and on the return journey to England he read a history of Venice in which he came across the name 'Degna', which he chose for his new daughter.

Proud though he was to be a father, the glamorous, itinerant lifestyle Marconi enjoyed so much, mingling on Atlantic liners with great singers, actors and actresses, and the newly emerging film stars, must have made home life seem dull. He and Bea had many rows, which began to worry her family. And he could not relax, for though he was now universally regarded in the popular imagination as 'the inventor of wireless', he knew very well that the technology was changing fast, and that at any time he might be toppled from his pedestal. The huckster Abraham White was still in business with his newly formed company, United Wireless, which was fitting out ships with shoddy equipment derived from patents other than Marconi's. And Lee de Forest was making great claims for his audion – which looked suspiciously like a copy of a Marconi Company invention – and was promising to transmit speech and music to audiences in America.

Although the Marconi Company's own Sir Ambrose Fleming had invented the first 'valve', the very piece of equipment that would bring the era of broadcasting in wireless, it was treated in the first years as a mere novelty. Marconi himself remained wedded to the technology he trusted, adapting it gradually but always making sure that it worked as advertised, and was therefore attractive to shipping lines, who put greater store by reliability than anything else. The Marconi Company now had several training schools for young operators,* who supplied a complete service to those who bought its equipment. They fitted out the wireless cabins on the ships, taking the rank of junior officers, and despite their age were soon afforded considerable respect as the guardians of a ship's safety at sea. When William Preece had sung Marconi's praises in his Toynbee Hall lecture of 1896, and said that the magic boxes demonstrated that night would ensure that sailors would soon have a 'new sense and a new friend', he had, for once, grasped the true potential of wireless telegraphy in the first two decades of its existence. Marconi might have grander ambitions to link the whole world with wireless stations, but it was out on the Atlantic that he achieved his first great life-saving triumphs.

---

* The first training school was opened at Frinton in Essex in 1901, later moving to Chelmsford. A second school opened in Liverpool, also in 1901, and another at Marconi House in London in 1912. There were also schools in New York and Madrid. The course took six weeks.

34

Wireless to the Rescue

At first light on a bitter January morning in 1909 a thick sea mist blanketed the treacherous waters of Nantucket shoals, which lie in the busy sea lanes off the coast of Massachusetts. On a wild headland of Nantucket Island, once celebrated for its whaling fleet and immortalised in Herman Melville's epic *Moby-Dick*, the wireless operator in the Siasconset station had fallen asleep and the coal on his fire had nearly all burned away. Jack Irwin had had a quiet night, with few messages coming in from Atlantic liners. Even the big ships equipped with wireless had just one operator on board, who would catch what sleep he could with one ear always cocked towards the receiver, which might at any time tap out an important message.

It was the cold which woke Irwin, who roused himself to rebuild his fire. As he was shovelling on the coal he heard his receiver come alive. He put on his headphones and picked up the faint but thrilling sequence which spelled out 'CQD'. It was the distress call of a Marconi operator: 'CQ' was the general call-sign 'Seek you', and the 'D' stood for distress.

Irwin tapped out his call-sign, 'MSC', then took down the incoming message, translating from Morse: 'Republic wrecked. Stand by for Captain's Message.' He knew that the *Republic*, a White Star liner, could not be far offshore, for its wireless had a range of barely seventy miles. Irwin relayed the distress signal, and told the *Republic* he was calling for tugs to go to her rescue. Then the *Republic's* wireless died.

That any message had got through at all was a small miracle, for the liner's lone operator, twenty-six-year-old Jack Binns, an experienced Marconi man, had very nearly lost his life when the ship, creeping through the dense mist with its automatic foghorn sounding, had been halted with a tremendous shudder. Binns had been asleep, and at first had no idea what had happened. He dashed from his bunk to the wireless cabin, where he saw the bows of another ship smash into the side of the *Republic*, wrecking cabins on the upper deck, slicing into the engine room and tearing away part of the wireless cabin.

Binns threw on some clothes and went straight to the Morse key to hammer out the 'CQD'. For a few moments the ship's electricity, which powered the wireless station, remained on, then everything went black. Binns rigged up the emergency batteries and checked that the aerial was still intact. He could not use the phone to the bridge, which was down, but the ship's commander, Captain Inman Sealby, gave instructions for Binns to send the message 'Republic rammed by unknown steamer 175 miles east Ambrose Light. Lat. 40.7, lon. 70. No danger to lives.' This was optimistic, for the *Republic* was clearly sinking.

Jack Irwin's general distress signal from Siasconset was picked up by two liners, of which the *Baltic* was closest to the *Republic*. Its operator soon established contact with Binns. He told them to hurry.

The *Republic* had left New York at 5.30 the previous afternoon, with 1600 passengers, including many well-to-do Americans embarking on a European tour. It had been rammed by a much smaller liner, the *Florida*, which carried two thousand passengers, many of them refugees from Messina in Italy, where they had been made homeless by a recent earthquake. Powerless, and shipping water all the time, the *Republic* was dragged around by the strong currents. Captain Sealby asked for permission to transfer all his passengers to the already heavily loaded *Florida*. This was done with the *Republic*'s lifeboats, each of which had to make several journeys as there were not enough for all the passengers on board.

Meanwhile, Binns stayed at his post, calling the *Baltic* and guiding it towards them in the dense sea mist. The *Republic* was sinking at the rate of about one foot every hour as its 'watertight' compartments filled. The *Florida* stood by, badly damaged and dangerously overloaded. Captain Sealby relied on the trained ears of Binns to estimate how far away the *Baltic* was as its wireless signals became perceptibly stronger, little by little. Around noon it was thought the majestic liner was just ten miles off, but the fog had thickened, and it had had to reduce speed because of the very real danger that it would suddenly emerge from the gloom and ram the *Republic*. Many other ships had by now picked up the distress signals, either from the *Republic*'s own wireless or from the Siasconset station.

All through the afternoon Binns kept at the Morse key, drawing the *Baltic* in until Captain Sealby thought it might be within earshot. Rockets and flares were fired by both ships, but there was no contact. The *Republic* let off its last direction-finding bomb, but the *Baltic*'s crew did not hear it. Finally, at about six o'clock in the evening, Binns asked the *Baltic* to detonate its last bomb. On the *Republic* Captain Sealby, the seven crew members who had remained on board, and Binns stood in a circle and listened intently. Binns heard a faint explosion, and an officer confirmed it. Binns, who had been at the Morse key almost continuously for fifteen hours, returned to the wreck of the wireless cabin and sent the *Baltic* their bearings. He was still in the cabin when he heard the *Baltic*'s foghorn, and then loud cheering as it pulled alongside them with its passengers leaning over the side, waving.

The captain of the *Baltic* was asked to find the *Florida* and to take all its passengers aboard for safety, leaving the *Republic* wallowing and powerless in the swell. Nearly four thousand passengers were taken from the *Florida* in lifeboats and put aboard the *Baltic*, which was a huge liner and capable of carrying many more than its legal limit. Recalling this great drama, Jack Binns said later:

. . . when daylight broke the next morning, Sunday, there was one of the greatest concourses of ships ever seen on the seas. Everywhere as far as the eye could see were ships. Every liner and every cargo boat equipped with wireless that happened to be within a three hundred mile radius of the disaster, overhearing the exchange of messages between the *Baltic* and *Republic* had gathered around and stood by ready to be of whatever assistance they could. It was a fine testimonial to the value of wireless. Shortly after daybreak the *Baltic* proceeded to New York and the *Florida* also proceeded at slow speed, convoyed by two or three other ships that were standing by. And then relief ships cared for the badly damaged *Republic*.

Jack Binns was the archetypal Marconi operator. Born in Lincolnshire in 1884, he had shown an interest in electrical science and had got a place at a technical school run by the Great Eastern Railway. He learned Morse code and the job of the telegraph operator, and was an ideal candidate for the new world of wireless telegraphy. After he joined the Marconi Company he was sent to Belgium, and became an operator on a German ship fitted with Marconi wireless. He might have stayed there had Germany, which was still campaigning to bring the Marconi Company into line, not made the decision to dismiss all foreign operators from its ships. Germany had established at the Berlin Radiotelegraphic Conference of 1906 that the international distress signal should be its own: the letters 'SOS', which in Morse is three dots, three dashes, and three dots. Marconi operators chose to ignore this ruling, and continued to use their own distress call, 'CQD'. Out on the Atlantic the chances were that wireless operators would know each other personally, and Marconi men could behave as if there were no international rules or laws governing their behaviour.

While it was being towed in, the badly damaged *Republic* sank, with the captain and first officer still aboard. They were rescued after a search, and Jack Binns finally steamed back to New York.

THE TRANSATLANTIC TIMES.

| VOLUME I. | NUMBER I. | BULLETINS |

THE TRANSATLANTIC TIMES

Published on board the "*ST PAUL*," at Sea, *en route* for England, November 15th, 1899.

One Dollar per Copy in aid of the Seamen's Fund.

Mr W W Bradfield, Editor-in-Chief. Mr T Bowden, Assistant Editor. Miss J B Holman, Treasurer. Mr H H McClure, Managing Editor.

Through the courtesy of Mr G Marconi, the passengers on board the "St Paul," are accorded a rare privilege, that of receiving news several hours before landing. Mr Marconi and his assistants have arranged for work the apparatus used in reporting the Yacht Race in New York, and are now receiving dispatches from their station at the Needles. War news from South Africa and home messages from London and Paris are being received.

The most important dispatches are published on the opposite page. As all know, this is the first time that such a venture as this has been undertaken. A Newspaper published at Sea with Wireless Telegraph messages received and printed on a ship going twenty knots an hour!

This is the 52nd voyage eastward of the "St Paul." There are 375 passengers on board, counting the destitute, guished and extinguished.

The days' runs have been as follows:—

Nov. 9th	435
,, 10th	436
,, 11th	425
,, 12th	424
,, 13th	431
,, 14th	414
,, 15th	412

97 miles to Needles at 12 o'clock, Nov. 15th.

BULLETINS

1.50 p.m. . . . . —. First Signal received, 66 miles from Needles.

2·40 " Was that you "St. Paul"? 50 miles from Needles.

2·50 Hurrah! Welcome Home! Where are you?

3·30 40 miles. Ladysmith, Kimberley and Mafeking holding out well. No big battle. 15,000 men recently landed.

3·4- " At Ladysmith no more killed. Bombardment at Kimberley effected the destruction of ONE TIN POT. It was auctioned for £200 It is felt that period of anxiety and strain is over, and that our turn has come."

4·00 Sorry to say the U.S.A. Cruiser "Charleston" is lost. All hands saved

The thanks of the Editors are given to Captain Jamison, who grants us the privilege of this room.

The first news-sheet ever produced on an ocean liner with reports transmitted from the shore by wireless. Returning from New York in 1899 on the SS *St Paul*, Marconi persuaded the captain to allow him to use the print room to produce the *Transatlantic Times*, with news sent from the Royal Needles Hotel. Passengers paid $1 a copy, which was donated to a fund for seamen.

A love letter from the American heiress Josephine Holman, who Marconi met on the liner *St Paul* in 1899. The shipboard romance had led to a secret engagement. Back home in America, Josephine wrote some of her letters to Marconi in Morse code in case her mother, who did not know of her engagement, found them.

Above The wild and desperate scene of Marconi's greatest achievement, which astonished the world and was greeted with scepticism by many scientists. Here he watches as helpers attempt to get a huge kite airborne at St John's, Newfoundland, in December 1901, to carry an aerial with which he hoped to pick up a signal from Cornwall on the other side of the Atlantic.

Right The proof Marconi needed to dispel doubts that he could send wireless signals more than two thousand miles: a chart, signed by the captain and first mate of the SS *Philadelphia*, confirming the distances Morse messages were received as the liner sailed from Southampton to New York in February 1902.

Above right A 'magnetic detector' devised by Marconi in 1902 to replace his earlier receiver, the coherer. Though it looks makeshift in its cigar-box housing, and included a motor cannibalised from an Edison phonograph and coils made of florists' wire, the 'maggie', as wireless operators called it with affection, served Marconi well for many years, and was the prototype of the receiver used on the *Titanic*.

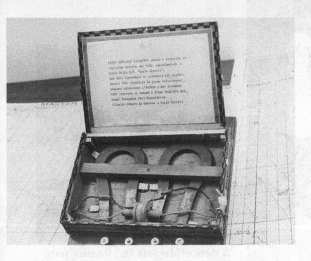

A favoured, adopted son of the British upper crust, Marconi is here decked out in some kind of ceremonial regalia as he accompanies his friend Sir Henneker Heaton MP to a bash at Westminster to celebrate Edward VII's coronation in 1902.

A thorn in Marconi's flesh for many years, the boastful and eccentric American Lee de Forest, posing with his claim to fame, the 'audion', an early wireless valve, and a later, miniaturised model. De Forest styled himself 'the Father of Radio', but all his claims to inventive originality were hotly disputed.

American wireless pioneers and their unscrupulous backers were keen to convince a gullible public that they had beaten Marconi at his own game. This exhibit, put on at the St Louis World's Fair in 1904 by Abraham White, the fraudster who funded Lee de Forest, claims, with some justification, two 'firsts': the Morse-code car-phone, and the use of wireless in war, to report on naval battles between Russia and Japan.

Left A dramatic illustration of how Marconi's practical but anachronistic technology grew to outlandish proportions. These are the gigantic power packs of his transmitter at Clifden, on the west coast of Ireland, built before he discovered there were more efficient ways to send signals across the Atlantic.

Below The final scene of what the French news-papers called 'Le Match Drew–Crippen', in which Inspector Drew of Scotland Yard chased Hawley Harvey Crippen, wanted for the murder of his wife, across the Atlantic. The captain of the liner *Montrose*, on which Crippen hoped to escape with his lover Ethel Le Neve, saw through their disguises and alerted Inspector Drew by wireless.

When the *Titanic* began to sink in the early hours of 15 April 1912, the shocking news was relayed across the Atlantic by Marconi operators. Tragically, this message from the liner *Virginian* to the *Californian*, the ship nearest the *Titanic* when it went down, was to no avail. It was sent at 4 a.m., nearly two hours after the *Titanic* sank, but was not picked up by the *Californian*'s lone wireless operator, who was asleep.

Marconi was acknowledged worldwide as the hero of the *Titanic* disaster, for without his invention on board, and the wireless distress signals sent out, the fate of the ship would have remained a mystery and nobody would have survived. Many perished because of a lack of lifeboats, a point made in this American cartoon, which has Marconi telling King Neptune he can 'beat him anytime' – once the shipping lines provide the rescue boats.

— Marconi to father Neptune —
I can beat you out any time
if you will only give me a few more of these life boats —

Grant Wright N.Y.
Apr. 17. 1912.

A family snapshot of Marconi with his wife Bea and their three children Degna (left), Giulio and Gioia. Taken around 1920, this was not a picture of domestic bliss, for Bea was not prepared to endure her husband's long absences and infidelities for much longer; they were divorced three years later, when Bea herself had found a lover whom she wanted to marry.

Oliver Lodge, an eminent English physicist and wireless pioneer, making a broadcast in one of the first radio studios in the early 1920s. Though Lodge was sometimes bitter about Marconi's commercial success, he showed a greater interest himself in spiritualist séances and discovering a proof of life after death than in developing a workable wireless system.

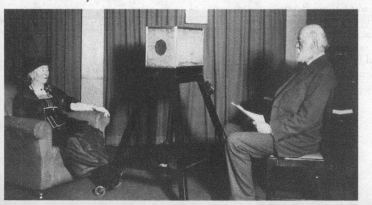

Above The mobile Marconi wireless kit proudly displayed by British soldiers in the First World War. Wireless was used in the trenches, where a Marconi engineer developed direction-finding techniques, but it had its greatest impact in the decoding of German naval messages picked up by listening stations on the English coast.

Right Marconi with his second wife Cristina Bezzi-Scali enjoying the sunshine of his fame on the yacht *Elettra*, which he bought and had fitted out as a floating home and laboratory in 1920.

It was then that his real nightmare began. At the White Star pier a group of sailors and stewards hoisted Binns and Captain Sealby onto their shoulders and carried them through the cheering crowd, amidst a fanfare of trumpets. The *New York Times* and all the other newspapers and magazines had followed the drama, and Jack Binns was treated like a modern pop star. He did his best to shun this unwanted publicity, but the press would not leave him alone. The Marconi Company exploited his celebrity to the full, producing publicity cards with a photograph of Binns, captioned 'Wireless hero' and decorated with their own 'CQD'.

On their way into New York from Nantucket the passengers on the *Baltic* made a collection, the money from which was used to strike commemorative coins of the event. All the crew members of the *Baltic*, *Republic* and *Florida* received medals, but four were struck in gold, one each for the three captains and Jack Binns. On one side of the medal was the distress call 'CQD' above a picture of a ship with 'SS *Republic*' below, and on the reverse was a citation commemorating the rescue.

Binns was a reluctant hero, but Guglielmo Marconi basked more happily in the reflected glory of the story: it was his invention that had saved nearly four thousand lives. The only casualties had been three passengers on the *Republic*, who were killed when the *Florida* rammed into their cabins. The magazine *Harper's* wrote up the drama of the rescue as if wireless had only just been invented: 'What a wonder-tale it is and how deeply moving – the cry for help thrown out into the air from a mast-tip, and caught, a hundred miles or more away . . . It is a new story; there never was one quite like it before.'

Jack Binns could not wait to get away from New York and back to work. But when he arrived back in Liverpool on the *Baltic* there were again huge crowds, who saw him reunited with his girlfriend. In his home in Peterborough he had to endure a mayoral reception. Finally, Marconi himself presented him with a commemorative watch.

From then onwards the Marconi wireless operator was no longer an intriguing but minor figure who tapped out inconsequential messages for first-class passengers: he was the saviour on the high seas. With his horror of publicity and entirely genuine puzzlement at the adulation he received, Jack Binns proved himself a true heir to the young Marconi, modestly brushing aside the plaudits, insisting that he was only doing his job. None of the boasts of Marconi's rivals could compare with the triumph of this rescue at sea: it was not long before he would be awarded the highest of international honours for his achievements, and it was probably the rescue of the *Republic* which tipped the balance in his favour.

❊❊❊❊❊

35

Dynamite for Marconi

On 10 December 1896, just two days before Guglielmo Marconi's demonstration of his magic boxes at Toynbee Hall in London, an inventor with more than three hundred patents to his name died in San Remo, Italy. Alfred Nobel was both an immensely wealthy and an immensely cultured man, a poet who was fluent in Russian, French, English and German as well as his native Swedish. He was born in Stockholm in 1833, the son of Immanuel Nobel, an enterprising builder. The family were bankrupted by an unfortunate accident in which a consignment of construction materials was lost, and Immanuel took them first to Finland and then to Russia, where his wife supported them by running a grocery store. Immanuel had been experimenting with explosives to blast rock for his construction work, and in time he was awarded a contract to provide the Russian army with equipment, including explosives for a crude form of naval mine. The family settled in St Petersburg, where Immanuel prospered and arranged for his children to be educated by private tutors.

The young Alfred Nobel, whose fondness for literature was regarded by his father as unfortunate, was despatched to various parts of the world to work with chemical firms and academics, and finally settled in Paris. There he learned about the highly volatile explosive nitro-glycerine, and began to experiment with ways of making it serviceable. Meanwhile his father's Russian business collapsed and the family fortune was once again in jeopardy until two of Alfred's brothers revived it by developing the Russian oil industry. Alfred returned to Sweden, where he continued to experiment with nitro-glycerine. It was a very dangerous substance: one of his brothers was killed along with other workers in an explosion before Alfred had found a way of making it safe, producing nitroglycerine in solid sticks which could only be detonated with a blasting cap. He called it 'dynamite', and patented it in 1867. Dynamite made him his fortune.

At the age of forty-three, Alfred Nobel was still single. The French novelist Victor Hugo described him as 'Europe's richest vagabond'. Desperate for a relationship, he put a lonely hearts advertisement in a newspaper: 'Wealthy, highly educated gentleman seeks lady of mature age, versed in languages, as secretary and supervisor of household.' The post went to Countess Bertha Kinsky, an Austrian aristocrat who stayed with Nobel for only two months but remained a friend for the rest of his life. She married a fellow Austrian, Count Arthur von Sutter, and became an activist in the European peace movement towards the end of the nineteenth century. Nobel, though famous for his invention of dynamite, became a supporter of the peace movement – explosives were more widely used for blasting rock for roads and mines than in warfare. He always retained his interest in literature and poetry.

When he died, aged sixty-three, Nobel was unmarried and childless. His family were startled to find that he had left most of his fortune in the trusteeship of two engineers, with instructions that they should award annual prizes for the best work anywhere in the world in the fields of physics, chemistry, physiology or medicine, literature and peace. By 1901 the Nobel Foundation was established, and was deliberating on who should receive the first awards. Scientists around the globe were asked to put names forward. One of those consulted was Ambrose Fleming, and it occurred to him straight away that Marconi should be a contender. He wrote to him: 'I do not know whether you have ever heard of the Nobel Prizes of the Swedish Academy. Mr Nobel was the inventor of dynamite and he left an enormous fortune to the Swedish Academy for funding Institutes and awarding prizes to inventors . . . One of these is for physics or inventions in physics . . . It occurs to me that your name should be put forward . . . and I shall be very pleased to nominate you.'

That first-ever Nobel Prize for physics in fact went to Wilhelm Conrad Roentgen for his discovery of X-rays, which had already had a profound influence on medicine. Wireless was still very new, and Marconi had not yet spanned the Atlantic at the time the first contenders were being considered. For the next seven years his name was always up for consideration by the Swedish Academy, but the award was given to others. Marconi had no expectation of winning it, because he was always conscious of the fact that he was not, technically speaking, a proper scientist with qualifications, and had only honorary academic posts awarded in recognition of his practical achievements.

After the excitement of the great sea rescue in January 1909, Marconi settled back into his work routine, enjoying his frequent trips across the Atlantic in the sumptuous saloons of luxury liners. Beatrice, his young daughter Degna and his mother would still sometimes stay at the Haven Hotel in Poole, where in his few spare moments the faithful George Kemp would try to keep them amused. On occasion, following forlornly in her husband's footsteps, Beatrice would travel to the remote wireless stations. The largest of these, at Clifden, was in her native Ireland, and it was while she was staying there in the autumn of 1909 that she discovered she was pregnant once again. Marconi was on a liner returning from America, and in her youthful eagerness Beatrice thought it would be exciting to break the news to him as soon as possible, not by telegraph but face to face before his ship arrived. A tug was due to take supplies to the liner as it approached the Irish coast, and she persuaded the skipper to take her aboard, and to let the captain know that an unexpected passenger should be added to the distinguished list in first class. It should have been a wildly romantic meeting. Beatrice found Marconi in the company of the great tenor Enrico Caruso and a party of glamorous actresses. The sight of his wife broke the spell. Instead of greeting her warmly, Marconi was angry at this intrusion on his other life. Beatrice was heartbroken, and locked herself in his cabin until the liner docked at Liverpool, refusing all his apologies.

It took a long while for the wound to heal, but by December 1909 they were reconciled. Marconi was away again, but he had read about an exciting rumour. He wrote to Beatrice: 'Some of the papers have said that I have got the Nobel Prize of £8,000. It rather makes one's mouth water to think about it just now, but I suppose it's not true.'

The reports turned out to be correct: Marconi was to share the 1909 Nobel Prize for physics with Karl Ferdinand Braun, the Principal of the Physics Institute at the University of Strasbourg. Braun was hardly known to the public, though he had invented the cathode ray tube, which was to become an essential component of television. He had also studied, both theoretically and practically, many of the same problems as Marconi, although the rivalry between Germany and the Marconi Company had ensured that they knew little of each other's work. When the two met for the first time in Stockholm both were in a state of pleasant surprise that they had received a share of the award: Marconi because he was not really a scientist, and Braun because he imagined his work was virtually unknown outside German academic circles. Perhaps because he was so conscious of his lack of academic credentials, Marconi's Nobel speech was highly technical, a detailed account with diagrams of all his experiments going back to 1894. He delivered it in English rather than Italian, explaining at the end that he felt this was a more widely understood language.

Marconi was frank about the fact that he did not understand how his signals had spanned the Atlantic, though he ran through the existing theories about waves bending and the effects of the earth's 'magnetic fields'. He made no mention of Oliver Heaviside, who was more than once put up for the Nobel Prize himself, or of Heaviside's theory about waves being reflected by the ionosphere.

There can have been few successful inventors with as little theoretical understanding of their own achievements as Marconi. But that did not matter much to him, or to his companies, which were now the world's foremost makers and providers of wireless telegraphy. If an Atlantic liner had a wireless cabin, the equipment would almost certainly be Marconi's, and the young man operating it one of his employees.

Before they left Stockholm, Marconi and Beatrice, now four months pregnant, were invited to a reception at the Royal Palace along with the other Nobel laureates. Beatrice and her sister Lilah were excited to discover that their old nursemaid Agnes from Dromoland was now looking after the sons of the Crown Prince and Princess of Sweden. Agnes, they found, had portraits of the O'Briens all over her room. Beatrice collected the autographs of other famous guests, and she and Marconi appear to have enjoyed their time together in Sweden with no jealous fits or marital upsets.

Marconi very much hoped that the child Beatrice was carrying would be a boy, and when they returned from Stockholm he persuaded her to go to Italy so that the child would be born an Italian national. Beatrice dutifully complied, though she would have been happier in England.

When, on 21 May 1910, Beatrice gave birth to a son in Bologna, Marconi was once again crossing the Atlantic. He had travelled so frequently that Beatrice did not know which ship he was on. To break the news, she sent a message addressed simply to 'Marconi – Atlantic'. Passed from one operator to another, it found him with very little delay. By this time there was no escaping Marconi's wireless signals, which formed an invisible web lying over thousands of square miles of ocean.

36

Le Match Dew–Crippen

The electronic crackle of a Marconi wireless aerial had become a familiar feature of the ocean liners that crossed the Atlantic in the years just before the First World War. For the passengers, most of whom were on the westbound voyage from Europe to Canada or the United States, the sound of the sparks flying as messages were fired from the ship's aerial was both magical and reassuring. Few of those aboard could afford to send greetings to friends and relatives ashore,* but the knowledge that their ship was not cut off from the rest of the world when the seas ran high, or they skirted close to a floe of icebergs, or the fog came down, was a comfort. While they took a stroll on deck they might wonder what news was flying between the wireless cabin on their ship and the young Marconi men aboard other liners out in the Atlantic or in the many shore stations that lined the British and North American coastlines.

Among the passengers on the *Montrose* of the Canadian Pacific Line, crossing in July 1910 from Antwerp to Quebec, was a small, retiring man who often paced the deck. Hearing the sparking of the aerial, he would remark by way of idle conversation to

---

* On the *Titanic* in 1912 the cost was twelve shillings and sixpence for the first ten words, then nine pence per word after that (at today's prices, £30 for the first ten words and nearly £2 a word thereafter).

the ship's captain, Harold Kendall, what a wonderful invention Marconi's wireless was. Kendall, who had been on the first-ever ship to carry commercial wireless equipment, would nod in agreement as he studied the passenger, Mr Robinson, closely. He noted that Robinson had shaved off his moustache and was growing a beard, and that he had spectacle marks on the bridge of his nose, although Kendall never saw him with glasses on aboard the *Montrose*. Robinson nevertheless read a great deal, and the captain made a note of his choice of books.

Mr Robinson was travelling with his son, who appeared to be in his early twenties. A strange boy, Captain Kendall thought, whose hat appeared to be stuffed with paper to make it fit, and whose loose-fitting trousers were held together at the back with safety pins. He ate his meals with the delicacy of a lady, and seemed on occasion to hold his father's hand rather tightly. Mr Robinson explained that his son was not in the best of health, and that they were en route to California for the climate. They had been to America before, as well as to Canada, where they were now headed with the plan of taking a train to the west coast.

Before the *Montrose* left Antwerp, newspapers in Britain, and in fact all over Europe and North America, had carried stories of the hunt for a murderer, an American called Hawley Harvey Crippen who had lived in London for a number of years with his second wife, a minor music hall performer with the stage name of Belle Elmore. Though Crippen was a quiet, mild-mannered little man, he appeared to be devoted to his ebullient wife, who enjoyed many friendships in the theatrical world. They often had guests at their home at Hilldrop Crescent in North London. Then, quite suddenly, the vivacious Belle Elmore had disappeared. She had last been seen on 1 February 1910. Crippen told her friends she was unwell, and had gone back to America. On 26 March he had a notice put in the newspapers that she had died in California. At the same time it was apparent that Crippen was living with his secretary Ethel Le Neve, who was sometimes seen wearing Belle's jewellery.

Friends of Belle, suspicious about her mysterious disappearance and her husband's obvious attachment to another woman, began to make enquiries, and found that there was no record of her death in California. After more than one approach to Scotland Yard, Detective Inspector Walter Dew was put on the case, and finally interviewed Crippen and Ethel Le Neve. Crippen, who called himself 'Dr' Crippen and was a dispenser of various medicines for the cure of deafness and other disorders, seemed a respectable sort of chap. After several conversations he confessed on 8 July that he had lied about his wife's death. She had in fact run off with her lover, he said, and he had made up the story of her illness to avoid social embarrassment. Crippen convinced Inspector Dew that nothing suspicious had happened, and Dew would perhaps have closed the case had not Crippen and Ethel Le Neve suddenly disappeared on 9 July, when they were due to give him a final interview. Two days later police searched the house at Hilldrop Crescent, and noticed a loose brick in the basement. When they began to excavate they found, covered in lime, some body parts of an unidentifiable woman. The hue and cry for Crippen and his lover was on. Newspapers around the world published his photograph and the gruesome details of the murder. Many innocent little men were hauled into police stations from southern France to northern Belgium, but none of them proved to be the notorious Dr Crippen.

There were newspapers aboard the *Montrose* with a photograph of Crippen, showing him bespectacled and sporting a droopy moustache. Quietly Captain Kendall had them all collected and put away. Very early on in the voyage he had become convinced that Mr and Master Robinson were none other than the murderer Hawley Crippen and his lover Ethel Le Neve. Kendall could have arrested the pair himself, but he preferred instead to keep a watch on them. He was concerned that they should not become suspicious, as he had noticed that 'Robinson' had a pistol in his pocket, and feared that he might kill his 'son' and himself if he were confronted. The maximum range of the Marconi wireless cabin aboard the *Montrose* was less than 150 miles, but two days into the voyage, as the ship emerged from the English Channel, she was still in touch with Marconi shore stations. Taking only the Marconi operator and his senior officers into his confidence, Captain Kendall sent a message to be relayed to Scotland Yard that he believed Dr Crippen and Ethel Le Neve were on board.

Inspector Dew had no doubt that it was Crippen on the *Montrose*, and he was anxious to catch him before he had any chance of slipping away in Canada. A check with the shipping lines suggested he might just be able to make it. The *Laurentic*, which was about to sail from Liverpool, was faster than the *Montrose*, and should overhaul it at the mouth of the St Lawrence. Dew took a train to Liverpool, and was in pursuit within hours. The newspapers got hold of the story almost immediately, and provided their readers with a day-by-day account of the positions of the two ships, many of them publishing maps showing the *Montrose* and the *Laurentic* like two snails in a race across the Atlantic. The reading public were entertained daily with stories about Dr Crippen and his lover: how he had bought a poison, where he had acquired boy's clothing, and his flight from Antwerp on the *Montrose* under an assumed name. In France what the papers called '*le mystère de Londres*', and then '*le match Dew–Crippen*', caused particular excitement.

American newspapers brought out special editions as each new piece of information on the race between the *Laurentic* and the *Montrose* was relayed by wireless. Captain Kendall entered into the spirit of the chase with an enthusiasm not quite seemly in a British officer. When his ship was out of range of the British coast he managed to contact the Montreal correspondent of the London *Daily Mail*, and his detailed account of his observations of Dr Crippen and Ethel Le Neve was cabled back, to be printed in full. The whole world had every detail of the pursuit, but Crippen and his lover – and the other passengers on board – remained blissfully ignorant of the messages flying above their heads in Morse code.

As the *Montrose* approached Father Point in the estuary of the

St Lawrence, Captain Kendall kept the Canadian Chief Constable up to date: 'Crippen is having breakfast. Suspects nothing. Your instructions carried out to the letter. Le Neve not up yet.' The *Daily Mail*'s special correspondent sent a report, datelined Sunday, 31 July, which began:

> A sharp, cold wind blew up from the east and with it the fog from the Atlantic. Four masts and a funnel loomed indistinctly away out on the waters . . . From the shadows of the wharf a skiff shot out and lost itself in the mist. The dismal horn of the steamer hooted, and the bell from the lighthouse buoy sent forth its message of guidance and assurance. In the skiff sat four sailormen – pea-jacketed, brass buttoned, visor capped officers of the pilot service. They rowed hard with grim determination in each stroke.

Only one of these men was a true river pilot. The others were police officers, including Inspector Dew of Scotland Yard, who was disguised in a captain's uniform. When they had hauled themselves aboard, Dew shook Captain Kendall's hand, and after a glance at 'Mr Robinson', muttered: 'That's my man!' Crippen did not hear this, but continued to pace the deck until he was confronted by Dew. He lost his composure as the Canadian Chief Constable put him under arrest. A gathering throng of excited passengers then heard a woman's scream: Dew had found Ethel Le Neve in her cabin, and she had become hysterical and then collapsed. She too was arrested, and both were taken back to London to stand trial at the Old Bailey.

For many years after the trial, the Dr Crippen case fascinated the public on both sides of the Atlantic. Alfred Hitchcock would weave elements of the story into several of his spine-chilling films. The motive for the murder was never established beyond doubt. One theory was that Crippen had no intention of killing his wife, and wanted merely to put a stop to her infidelity: the poison he gave her was recommended as a potion for quelling the libido of sexually excitable women. But the prosecution claimed that he had murdered her in order to get his hands on her jewellery, some of which he had pawned to pay for his flight with his lover Ethel Le Neve. Crippen always maintained that he was innocent, and impressed his prison warders with his kindly manner and personality.

In time the role played by Marconi wireless in Dr Crippen's arrest was forgotten, but in 1910 it was a sensation, yet another triumph for the new technology with which the public were becoming daily more familiar. The *Daily Mail*, ever attentive to the curiosity of its readers, commissioned a piece by W.W. Bradfield, now deputy managing director of the Marconi Company, to explain this marvel. It was headlined 'Pursuit by Wireless: Danger of Shipboard for Fugitives: Long Arm of the Law'. Under the subheading 'The Magic Cabin', Bradfield wrote:

> The little cabin, crowded with apparatus, is like a magician's cave. All kinds of appliances are stacked within it. Printed telegraph forms are scattered at one end of the instrument table. The operator on duty is wearing a telephone headgear, with receivers over his ears. Suddenly there comes to him a low musical note ... As he replies the cabin is transformed. A vivid electric spark throws a weird bluish light over the operator and his machinery ...

The French newspaper *Liberté* commented:

> Arrest by wireless telegraphy opens a new chapter in criminal history. Thanks to the invisible agent, we are able to follow every movement of Dr Crippen and his companion. It is admirable and it is terrible. The story of this sensational capture will rank with the greatest wonders of wireless telegraphy ... It has demonstrated that from one side of the Atlantic to the other a criminal lives in a cage of glass, where he is much more exposed to the eyes of the public than if he remained on land.

37

A Marriage on the Rocks

The romance of the Marconi wireless telegraph was sadly not reflected in its inventor's domestic life. While Dr Crippen was being pursued across the Atlantic, Guglielmo, Beatrice and their new son spent much of the summer at the Villa Griffone. With support from Marconi's mother, who usually took Beatrice's side in their disputes, they were persuaded not to end the marriage. But the situation was becoming impossible. Whenever Marconi and Beatrice spent any time together there were conflicts; then Marconi would be away for long periods without anything being resolved. Their reconciliations were always short-lived.

In September 1910 Marconi was at sea again, pursuing an ambitious scheme to set up a worldwide system of wireless communication for the Italian government. He sailed for South America with one of his more brilliant engineers, H.J. Round, and together they raised aerials from the liner with kites, picking up signals from the Irish station at Clifden from four thousand miles during the day, and seven thousand miles at night. Maximum distance, above all else, was still Marconi's ambition.

Beatrice had two younger sisters who were ready to 'come out' in the London Season of 1911, when debutantes attended a round of balls before being presented to King George V and Queen Mary at Buckingham Palace. She returned to London and rented a house in the grounds of Richmond Park to the south-west of London, joining in the fashionable round of shows and parties – the very life that Marconi was happy to enjoy himself, but attempted to deny Beatrice when he was not with her. At the very time the jury at the Old Bailey was finding Dr Crippen guilty of the murder of his wife, the marriage of the wireless wizard whose invention had caught the timid American was close to collapse. Crippen was sentenced to death, and hanged on 23 November 1910, consoled by the knowledge that his lover Ethel Le Neve had been acquitted. Under the assumed name Ethel Nelson she would live in Canada before returning to England, where she married an accountant who did not know her true identity. She was a grandmother when she died in 1967.

Beatrice made the most of her brief time in London in 1911. She told her daughter Degna:

> We wore suits of ribbed material with embroidered peek-a-boo muslin blouses with lace jabots and stifled boned collars. Waists were tiny. At Ascot – we all went to Ascot – gowns trailed on the ground and hats, Gainsborough style, were heavy with feathers and flowers and were secured to our heads with huge pins. I wonder, now, how we balanced them on our heads and why we did not blind our escorts with what amounted to swords piercing our hats. In the evening we wore diamond dog collars and sheath-like lamé or brocade gowns with long trains, which we had to hold up while we danced.

In August Marconi took Bea away from London society and back to the wilds of Cape Breton, where she had a miserable time, plagued by summer insects, seasickness and toothache. Then, as the battle between the two of them seemed certain to lead to a grim formal separation, Marconi was called away to Italy for his first taste of war. In a now-forgotten conflict, Italy declared war on Turkey on 29 September on the thin pretext of a dispute about territory in what is now Libya. A loyal Italian patriot, Marconi was eager to serve his country, and shipped aboard the *Pisa*, which policed the North African coastline, supporting ground troops and attacking Turkish positions. These included wireless stations built by the Marconi Company's great German rivals Telefunken – Marconi was on board when the *Pisa* destroyed a large installation at Derna. All the while Marconi continued to experiment with the Italian wireless equipment, and from the tone of his letters home obviously enjoyed himself. He wrote to Bea, who had travelled with him as far as Taranto in southern Italy: 'I am in the best of health and spirits. There is only one woman in the whole place, she is an old Arab. We have a splendid hospital ship here, beautifully equipped but with no nurses on board. They had to send them away for as they had no wounded to attend and nursing to do they flirted too much with the officers.'

Marconi's mother wrote to him on 12 December 1911:

> I am so glad to hear you are well and that the climate is so beautiful. I am sure it will do you good and I hope and pray that the Lord will keep you safe, free from all peril and danger. I can't help feeling anxious about you darling, for it is an awful dangerous country to be in at present, but I suppose you are always on the warship and I am glad they are all so kind and that you are well looked after. I was very glad to hear from dear Bea that she had accompanied you to Taranto and had seen you off, and that you were so happy going. She mentioned how charming the Admiral is, and the officers, and what a beautiful large cabin you have . . .

Free for a time from the pressures of company work, Marconi was in his element, enjoying once again the life of the dedicated inventor. He went ashore and tested mobile wireless equipment carried on camels, while aboard ship he had the admiring attentions of the officers, even if, as he explained to Bea, there were no romantic temptations like those offered by Atlantic liners. His company was at last prospering, paying a dividend to shareholders for the first time in 1910, and rapidly absorbing competitors in both

Britain and the United States. After a brief period in which Marconi ran the company himself, a new managing director, Godfrey Isaacs, a man of drive and vision recommended by Beatrice's family, had been appointed in 1910. Marconi's connections with the upper crust of English society and the business world remained crucial to his success as Isaacs embarked on a programme of establishing the Marconi companies as the foremost manufacturers and distributors of wireless telegraphy equipment in the world.

While Marconi was examining with an expert eye the Telefunken stations destroyed by Italian naval guns in North Africa, Isaacs was negotiating to put an end to the long-running rivalry with the German company. All Germany's efforts to force Marconi to communicate with ships carrying Telefunken equipment had failed. At the same time Marconi's companies around the world came up against what they called 'the Telefunken Wall' whenever they sought new customers. The licensee for Marconi patents in Germany, Austria and some other European countries was La Compagnie de Télégraphie Sans Fils, based in Brussels, and it faced collapse when the German company banned Marconi wireless from their own ships, and fought fiercely for European orders. Godfrey Isaacs managed to negotiate an end to the stalemate. A new company, the Deutsche Betriebsgesellschaft für Drahtlöse Telegraphie (DEBEG), was created in January 1911. The British and Belgian Marconi Companies had a 45 per cent stake, and Telefunken 55 per cent. All resources on ships and in shore stations were pooled, and Austrian wireless stations were soon absorbed.

At the same time Isaacs began to sue rival companies for patent infringements, and where this was not possible, to buy out the competition. Oliver Lodge, who in 1896 claimed to have sent wireless messages before Marconi, and who had developed the 'coherer' that Marconi copied and adapted, was a tough opponent. He had a patent on 'syntony' or tuning which rivalled Marconi's, and with his friend Alexander Muirhead he had formed his own wireless company which had won contracts from the British Army. After a court action in London the Marconi Company agreed to pay

Lodge £1000 a year for the seven years his patent had to run, bought out the Lodge-Muirhead Company to close it down, and hired Lodge as a consultant at £500 a year. In America, Isaacs took action against the huge but financially unsound United Wireless for patent infringement, at a time when its directors were being prosecuted for fraud. Very rapidly the American Marconi Company attacked and defeated its rivals, absorbing NESCO, the company which had backed Reginald Fessenden, as well as United Wireless. Finally Lee de Forest, who was again facing charges of fraud and was licking his wounds in California after a second disastrous marriage and the failure of his wireless telephony system, was put out of the picture. De Forest's backers had let him down, and though after 1918 he was to become a celebrated pioneer of radio and sound movies in Hollywood, in 1912 he was no rival for Marconi.

John Bottomley, managing director of the American Marconi Company, was able to report in 1912 that from meagre beginnings there was now $5 million available for investment. Marconi's commercial ambitions were at last being fulfilled. By now he had acquired a Rolls-Royce and a chauffeur, and for the first time there was the prospect of a settled home life after years of travel. Beatrice took a lease on a country house in the parish of Fawley in Hampshire called Eaglehurst. It was a return to the south coast of England where Marconi had first demonstrated that wireless worked, and was only a short distance from the Haven Hotel in Poole, where there was still a research station manned by Marconi engineers. Romantic and slightly run-down, Eaglehurst had manicured lawns which led down to the seashore, peacocks which drove Beatrice mad with their screeching, and grounds in which the children could play. Degna remembered:

> Eaglehurst stretched out wide and one storey high – all windows, vines and chimneys – to end in matching two storey octagonal wings with the crenellated tops that so obsessed a Gothic-minded generation of builders. What we youngsters enjoyed about Eaglehurst, besides our pony

cart and the sheltered, pebbly beach, was the tower. This was a curious eighteenth-century architectural folly, a narrow three-storey structure fitted with Regency bay-windows and crowned with a round three-storey turret surmounted with the inevitable battlements and a flag.

Marconi could have settled here, financially secure, resting on his laurels. Instead he was seldom at Eaglehurst, coming and going in his Rolls-Royce, often bringing one of his engineers to stay. Degna and her brother Giulio had a nursery at the top of a flight of stairs in one wing of the house, and their pet Pekinese, Manchu, kept guard on the landing outside, snapping and snarling fiercely at any unfamiliar visitor. It was Marconi's habit on his rare and unannounced visits to go straight to the nursery. Manchu saw him so infrequently that he did not regard him as one of the family, but would attack him, snapping at his trousers. After a time 'the little monster', as Marconi called him, was taken away, much to the children's sorrow.

The crumbling old towers of Eaglehurst were generally out of bounds to the children, but on special occasions they were allowed to climb them and enjoy the fine view over Southampton Water, where the great liners were led by tugs into the open sea for the crossing to Cherbourg in northern France, the first leg of the voyage to New York. From the decks high above the water the passengers were close enough to wave to them in their turret. It was from here that Beatrice and Degna waved goodbye one morning in April 1912 to a great liner they would never see again.

38

✤✤✤✦✦✦

Ice and the Ether

Over the centuries of transatlantic voyages, thousands of ships had limped into port with their hulls bearing gaping holes inflicted by icebergs. Of those which sank in mid-Atlantic with all hands after hitting icebergs there is no record before the days of wireless telegraphy. How many of the wrecks which lie on the bed of the North Atlantic were sunk by icebergs can never be known.

Every winter, great tongues of ice spread out over the sea off the southern coast of Greenland, covering more than seven hundred thousand square miles with compressed frozen water to an average depth of five thousand feet. When the spring comes the immense icebergs break away with a thunderous crash, and float south and west towards the coast of Labrador. They are scattered by winds and tides and currents, drifting with their peaks above the waves, nine-tenths of their treacherous bulk submerged and melting below the surface. Icebergs present the greatest threat to shipping in April, May and June, and to reduce the risk of running into them it was customary for transatlantic liners to take a more southerly course in those months.

There was no telling what each spring would bring. Some years there were few bergs in April, while in others an early thaw and northerly winds would drive large floes into the path of the liners. A constant lookout was kept when ships sailed into the danger zone, and all experienced navigators knew that some bergs were much harder to sight than others, especially at night, for they were not brilliant white but were darkened by the action of the sea as they melted. Sometimes they turned over, and the rocks and rubble embedded in them showed above the surface.

By the spring of 1912 all large passenger liners on the North Atlantic run were equipped with a wireless telegraphy cabin, the great majority of them manned by Marconi operators, and were able to warn each other of any danger from icebergs. Though the range of their transmitters was under two hundred miles, there were so many liners at sea at any time that wireless contact was almost always possible. The pursuit of Dr Crippen had demonstrated the way in which messages sent from ship to ship and passed on to shore stations which were linked by cable could keep everyone on the open sea in touch with each other and with the mainland. From 1906 the Marconi Company had been publishing charts which showed the date of departure of liners from European ports and their expected date of arrival in North America, crisscrossed with the departure and arrival times of ships sailing from west to east. The charts gave a rough idea of where any liner was likely to be in mid-ocean on a particular day.

Many of the young men who had taken courses at the Marconi training centres knew each other personally, and they would relieve the loneliness they sometimes experienced by chatting to each other in Morse code. They were at their keys for long hours, tapping out messages for first-class passengers, and when they did crawl into their cabins to snatch a few hours' sleep their ship would be out of touch. Before turning in they would generally sign off in Morse 'GNOM' – Good night old man.

Competition between shipping companies on the North Atlantic route remained fierce in the early 1900s, and the White Star Line, a British company which had been bought up by the American magnate John Pierpont Morgan, decided to make its mark with three huge liners that would dwarf the competition. They were built in the shipyards of Belfast, their massive hulls held together by millions of rivets hammered into place by skilled boilermakers.

The first of these great ships, the *Olympic*, was launched in 1911. One of the most experienced White Star Line captains, Edward Smith, was in command on her maiden voyage, and almost immediately had trouble manoeuvring a ship larger than any he had handled before. In a collision with a British naval cruiser the *Olympic* lost part of one of her giant propellers. Emergency repairs to the *Olympic* delayed the maiden voyage of the second huge White Star liner, the *Titanic*, which had completed its trials in 1911.

The Marconi Company had the job of fitting both ships with wireless equipment, including a trusty 'Maggie' detector, housed in a handsome mahogany box, and an exceptionally powerful spark transmitter working off five kilowatts of power, giving it a much greater range than any other liner at sea except for the *Olympic*. This transmitter was run on the *Titanic*'s generator, but in an emergency it could be operated with batteries.

It was decided that the amount of traffic likely to be generated by the wealthy passengers in first class warranted two operators who would work in shifts. Jack Phillips, the senior operator, was twenty-six years old, and had already served on a number of liners as well as at the Clifden station. He had many friends among the Marconi operators, one of whom, Harold Cottam, had recommended the twenty-two-year-old Harold Bride as Phillips's junior. Cottam had met him the year before when Bride went to a wireless office to enquire about the possibility of taking part in this exciting new line of work on transatlantic liners.

Harold Cottam had already signed on as the lone operator on a Cunard liner, the *Carpathia*. He was a year younger than Bride, but was a precocious operator – the youngest, at seventeen, ever to pass the examinations. Cottam was just one of a number of Marconi operators Phillips and Bride would know personally when they exchanged messages out on the Atlantic.

As an experienced operator, Phillips could tap out thirty-nine words a minute, whereas Bride's speed was only twenty-six. They agreed to divide up their shifts so that Phillips worked from 8 p.m.

until 2 a.m., and Bride from 2 a.m. to 8 a.m., the quieter time. During the day they would share the work as it came in. Their sleeping quarters were next to the wireless cabin or 'shack'. They could nip out on deck for a smoke, but there would be little time for relaxation.

At least they could share the workload. Harold Cottam on the *Carpathia*, like most other operators, had to grab some sleep when he could, leaving his cabin unattended. When Phillips and Bride were in range they would chat to Cottam if they had time. If the wireless traffic was heavy, operators could be very abrupt in their responses to idle enquiries. Three 'D's – 'dah-di-dit, dah-di-dit, dah-di-dit' – meant 'Shut up!' More forceful was 'dah-dah-dit, dah, di-di-di-dit', which spelled out 'GTH' – 'Go to hell.' Each ship had its own call-sign, so that operators knew instantly who was sending signals or trying to contact them. The *Titanic*'s was 'MGY': 'dah-dah, dah-dah-dit, dah-di-dah-dah'.

The *Titanic* was the most up-to-date ship afloat, offering electric lighting and heating in the first-class cabins – a luxury for many of even the wealthiest passengers. Everything was worked by electricity: the lifts, the kitchens, the refrigerators (another novelty for European travellers), all kinds of ventilation fans and the loading cranes. There was even an electrically operated 'camel' for toning up in the gymnasium. This was just the kind of fabulous floating fantasy world that Marconi enjoyed, and he was naturally invited by the White Star Line to sail as an honoured guest on the five-day maiden voyage across the Atlantic. Beatrice, Degna and Giulio were also booked on the *Titanic*, and looked forward to spending some time in New York while Marconi himself continued his company's demolition of its American opposition.

When the *Titanic*'s maiden voyage was delayed by the emergency repairs to the *Olympic*, Marconi had to cancel. After his spell with the Italian navy in the Mediterranean he was urgently needed by his American company, and he sailed on the *Lusitania* at the first opportunity. Then young Giulio fell ill early in April, and Bea decided to stay with the children at Eaglehurst. Sad and disappointed, she held Degna's hand tightly and waved from the turret of Eaglehurst to the hundreds of lucky passengers who leaned on the rail of the upper decks of the *Titanic* as the great liner eased out of Southampton Water on 10 April 1912.

39

❧❧❧❧

'It's a CQD, Old Man'

Late on the night of 14 April, Jack Phillips was at the Morse key in the *Titanic*'s wireless cabin. Harold Bride was fast asleep next door. They had had a miserable day: their equipment had broken down late the previous evening, and it had taken them seven hours to find the electrical fault and put it right. All the while, inconsequential messages paid for by the first-class passengers had piled up. There had been one or two reports from other ships as well, several warning of icebergs ahead, and these had been delivered to Captain Edward Smith. As arranged, Harold had been at the key from two o'clock that afternoon. Jack had been due to relieve him at eight, but sending messages continuously was strenuous work, and Jack could see that his young assistant was exhausted, so he took over at 7.30 and told Harold to get some sleep. For nearly four and a half hours Jack had been hammering away at the Morse key as the *Titanic* cruised at twenty-two knots through a sea which was miraculously calm for the North Atlantic. There was not even a swell.

At around midnight Harold woke up and shuffled into the wireless cabin in his pyjamas. His next shift did not begin for two hours, but he wanted to return the favour Jack had done him earlier, and offered to take over. Despite the reports of a large amount of ice in the sea, neither had any concern for the safety of the ship. There was nothing unusual about such warnings, and it was up to the captain to take whatever action he thought necessary. He might slow down or take a more southerly course, or, as many experienced captains did, keep straight on ahead but order a sharp lookout. At more or less the same time as Harold Bride was trying to persuade Jack Phillips to hand over the Morse key, their mutual friend Harold Cottam was getting ready for bed on the *Carpathia*, which was fifty-eight miles to the south-east of the *Titanic*, cruising at a sedate fourteen knots and heading clear of any iceberg danger.

Cottam had crossed the Atlantic before the *Titanic* left Southampton, and was now on his way back from New York to the Mediterranean. As a lone operator, he worked long hours: the previous night he had not got to bed until 3 a.m. Just before midnight on 14 April he listened in to Jack firing off his thirty-nine words a minute on the *Titanic*. Most of the messages were going to Cape Race, Newfoundland, where two Marconi stations had been built since the Anglo-American Cable Company's monopoly had expired. Cottam heard Jack sending off instructions for a dinner party in New York at which all kinds of luxuries were to be served.

The range of Cottam's transmitter was only about 150 miles, but he could still call up Cape Cod, from where he learned that a backlog of messages for the *Titanic* was piling up. In a comradely gesture he took down five or six of these, with the idea of passing them on to Jack when he was less busy. He then took off his headphones, slipped off his shoes and hung up his coat. Before shutting down he thought he would have a last listen-in to Jack, and perhaps send him a 'GNOM' – Good night old man. He could hear nothing from the *Titanic*, so he tapped out: 'I say, old man, do you know there is a batch of messages waiting for you at Cape Cod?'

Jack's transmissions had been silenced by the arrival in the wireless cabin of Captain Smith. As Harold Bride stood there in his pyjamas and Jack was wondering why the ship's engines had been stopped, the captain told them to stand by to send a message. They had hit an iceberg, and he was going to assess the damage. Neither Jack nor Harold had felt any impact, and they believed there was no cause for alarm. So did Captain Smith, until carpenters and others came running from the bowels of the ship to tell him that water was pouring in, and the engine room was flooding.

The *Titanic*'s designer, Thomas Andrews, was in a stateroom, noting in his meticulous way small improvements that could be made to this near-perfect liner. He went with the captain to view the damage, and realised straight away, as only the ship's designer could, that the punctures inflicted in the hull by a glancing blow from the iceberg were fatal. He said the ship would go down within two hours, and there was no way to save it.

Captain Smith went straight back to the wireless cabin and ordered Jack to send the distress signal. As a loyal Marconi operator he sent out a 'CQD', rather than the 'SOS' which had long been established as the official international distress call. The message was picked up more or less instantaneously by the station at Cape Race and ships nearby, though not by the nearest liner, the *Californian*, whose lone Marconi operator had already gone to bed. Harold Cottam on the *Carpathia* missed it too. But when he put his headphones back on to try to raise Phillips again with the news about the unanswered Cape Cod messages, he got a reply instantly. 'Come at once. We are sinking.' Cottam could not believe his ears. He replied: 'What's wrong? Should I tell my captain?' Jack's retort stunned him: 'Yes. It's a CQD, old man. We have hit a berg and we are sinking.' Jack gave an estimate of the *Titanic*'s longitude and latitude.

Henry Rostron, the forty-three-year-old captain of the *Carpathia*, whose crew had given him the affectionate nickname 'the Electric Spark', lived up to his reputation when he was woken in his cabin at 12.35 a.m. The *Carpathia* was immediately turned about and set on a course for the position given to Cottam by Jack Phillips, who was sending Morse distress signals continuously. Captain Rostron told Cottam to send a message to the *Titanic* that they were coming as quickly as possible, and would probably take four hours to arrive on the scene. He then called in his senior officers and told them to prepare the ship for the rescue. The chief engineer was ordered to cut the heating and to reserve all power for the engines, which were pushed to their limit, driving the *Carpathia* along at seventeen knots. The iron hull shuddered and the furnaces blazed down below. Three doctors aboard were detailed to take responsibility for the *Titanic*'s first-class, second-class and steerage passengers. Hot drinks were prepared, saloons converted into recovery rooms, and lifeboats got ready as the *Carpathia* ploughed through the night, keeping a constant watch for icebergs. They were soon amongst them, but Captain Rostron kept as near to his maximum speed as he dared, taking a zig-zag course as warnings were shouted to the bridge.

In the *Titanic*'s wireless cabin, Jack Phillips and Harold Bride remained cheerful and optimistic. Jack continued to send 'CQD's alerting all ships within range, and occasionally, almost as a joke, he tapped out an 'SOS'. Harold said later that he had suggested, tongue in cheek, that it might be their last chance to try out the official international distress call which, originating as it did in Germany, they preferred to ignore. They sent the first 'SOS' to their sister ship, the *Olympic*, which was five hundred miles away en route to Southampton.

From time to time Jack went out on deck to see what was happening, and at last it became clear to him that the *Titanic* was sinking. Just before 1.30 a.m. he sent a message that women and children were being put into lifeboats. A few minutes later he reported that the ship's engine rooms were flooding. Their distress calls were silenced for a few minutes when a crew member came into the wireless cabin and tried to wrest Jack's lifejacket from him. Harold got hold of the man and Jack punched him senseless.

As the *Titanic*'s power began to fail the *Carpathia* lost touch with her. The last message came at around 1.45 a.m., telling Cottam that the engine room was full to the boilers. Captain Smith put his head into the wireless cabin to say it was now every man for himself, but Jack and Harold kept on sending until all power was gone. Only then did they make a dash for the deck. It was 2.17 a.m., and the *Titanic* was on the point of going down. Harold jumped first, and managed to cling on to an overturned 'collapsible' lifeboat. Jack jumped, but Harold did not see where he landed.

At 2.45 a.m. the lookouts on the *Carpathia* saw what they believed to be a green flare in the distance, and the liner made towards it. But fifty minutes later they had seen no sign of a ship, nor any lifeboats. Dawn began to break around 4 a.m., and Captain Rostron cut the *Carpathia*'s engines, as he was certain he was now close to the scene of the disaster. Another green flare went up, and they saw their first lifeboat just a hundred yards away. Carefully Captain Rostron manoeuvred his ship to bring the lifeboat alongside, to the lee side of a light wind which was beginning to stir the sea. This was not easy, for they were in the midst of a huge floe of icebergs. Among the first twenty-five survivors taken in stunned silence on board the *Carpathia* was the *Titanic*'s fourth officer, twenty-eight-year-old Joseph Boxall, who confirmed what Rostron feared. The *Titanic* had disappeared at 2.30 a.m.

It took six hours for the *Carpathia* to find and rescue a little over seven hundred survivors.* A few of those taken aboard were dead of exposure, among them Jack Phillips. Harold Bride was just hanging on to life: he had spent hours lying on the upturned lifeboat. A doctor bandaged his feet, which were frozen from the icy water. Harold Cottam, who had been about to catch up on his sleep when the first 'CQD' sent him running in shirt and trousers to the bridge, was still at the Morse key eight and a half hours later. He was exhausted, but he still had work to do. When Captain Rostron had satisfied himself that there could be no more survivors, he had Cottam tap out messages to other ships on their way to the rescue that there was no need for them. The *Carpathia*, bobbing amongst the icebergs which in daylight presented an awesome spectacle, now had the grim task of searching for the dead. One officer counted no fewer than twenty-five bergs rising 150 to two

* There is no agreed figure for the survivors of the *Titanic*, as the exact number of people aboard has never been established. Estimates range from 701 to 713. One recent estimate is as follows: total passengers 1316, of whom 498 survived. Total crew 913, of whom 215 survived. This gives the figure of 713 survivors.

hundred feet above the water, and many other smaller 'bergy bits'.

Captain Rostron considered his options. It was suggested that the survivors might be taken on board the *Olympic*, which had turned about and was covering the five hundred miles to the scene of the disaster. But Rostron felt that a transfer of survivors at sea in small boats, which would have to be rowed amongst the icebergs, would be too traumatic. The *Carpathia* began cautiously to steer a way out of the ice floe, and headed back to New York.

40

After the Titanic

Harold Cottam's first messages telling of the disaster which had befallen the *Titanic* had to be relayed to coastal stations via the much more powerful transmitter on the *Olympic*. The world was desperate for news of the fate of the 'unsinkable' *Titanic*, but Cottam was near to collapse, and able to give only the barest details. When the *Olympic*'s operators asked if there was anything he wanted to say to New York, he tapped back: 'I have not eaten since 5.30 pm yesterday.' A list of survivors had been drawn up, and Cottam began to send their names, but as night fell he was so exhausted he could barely press the Morse key.

Two officers reported to Captain Rostron that Cottam was 'acting queer'. Rather than hold the world in suspense, Rostron asked Harold Bride if he would take over, so that Cottam could get some sleep. Bride, his feet heavily bandaged, was carried to the wireless cabin and began once again to work the key. Like Cottam, he was instructed to send only 'disaster related messages'. The two men took that to mean that they should concentrate on the long list of the names of the survivors, and should refuse requests from newspapers for more details of the tragedy. The silence of the *Carpathia* on the most sensational story for many years angered the American press, and rumours sprang up that Bride and Cottam had been instructed to hang on to their testimony so they could sell their accounts exclusively when they arrived in New York.

So urgent was the demand for news that US President William Taft sent out two US Navy scout cruisers which had a wireless range of 150 miles to intercept the *Carpathia* and gather some information ahead of the liner's arrival in New York. But the Navy operators were rebuffed by Bride and Cottam who, as Marconi men, regarded themselves as being in a different league when it came to tapping a Morse key. To confuse matters further, American Morse code at the time was not quite the same as the Continental version used by Marconi operators. For example, the three dashes of the letter 'O' in Marconi Morse spelled the number 'five' in US Morse, so that the 'SOS' signal used on the Great Lakes was 'S5S'. Harold Bride later told a *New York Times* reporter: 'The navy operators aboard the scout cruisers were a great nuisance. I advise them all to learn the Continental Morse and learn to speed up in it if they ever expect to be worth their salt.' He said the Navy operators were 'as slow as Christmas coming'.

To add to the frustration and confusion, the first reports of the fate of the *Titanic* had been garbled when the 'CQD's from Jack Phillips criss-crossed with wireless messages from other ships. Somehow newspapers got hold of a story that the damaged liner was being towed to Halifax, Nova Scotia, and that all the passengers were safe. It was never discovered how this cruel raising of hopes had occurred, but the finger of suspicion was pointed at the hundreds of wireless amateurs listening in to the unfolding drama from their attic rooms on the eastern seaboard of America. Questions would also be asked about the attitude of American Marconi operators, but Guglielmo Marconi himself was instantly identified as the saviour of the *Titanic* survivors. When the *Carpathia* was still two days out from New York he wrote to Bea: 'Everyone seems so grateful to wireless – I can't go about New York without being mobbed and cheered – worse than Italy.'

The intense public attention made Marconi reluctant to go down to the Cunard pier to meet the *Carpathia* when she arrived on the evening of 18 April. Instead he dined at the home of John Bottomley, managing director of the American Marconi Company.

But the *New York Times* would not let him be. A reporter knocked on the door of Bottomley's home at 254 West 132nd Street at 10.30 p.m. asking for written permission to interview Bride and Cottam. Marconi decided to go with the reporter, and they took the Ninth Avenue elevated railway down to 14th Street. An anxious crowd swarmed out of the train, carrying Marconi and the reporter with them. A cab was waiting to take them to the Cunard pier, where thousands had gathered to greet the *Carpathia*. In the crush Marconi was not recognised, and he had to talk his way past a series of guards before he could board the liner. He waited until all the survivors had been taken off, then made for the wireless cabin. Milking every drop of sentiment, the *Times* reporter who went with him described the scene:

> He almost ran forward and threw back the door of a tiny cabin. One lamp was burning in it. A young man's back was turned to him, and between two points of brass a blue flame leaped incessantly. Slowly the youth turned his head around, still working the key. The hair was long and black and the eyes in the semi-darkness were large – staringly large. The face was small and rather spiritual, one which might be expected in a painting. It was clear that from the first tragic moment the boy had known no relief. 'Hardly worth sending now, boy,' said Mr. Marconi, hoping to cause the youth to stop. 'But these poor people. They expect their messages to go.' And the operator, Harold Bride, caught Mr. Marconi's face and saw his hand extended. 'He recognised the man who discovered the wireless system, although he had never seen him before. He glanced from Mr. Marconi to a little picture above the wireless instrument. It was a picture of Marconi. They shook hands long and without saying a word. The boy's face changed in expression gradually. The strain of his long work was just beginning to break, and he smiled. 'You know, Mr. Marconi, Phillips is dead,' were his first

words. Mr. Marconi asked the operator how his feet were. Both were in bandages and he was working seated on the edge of his bed. A plate of food at his side told how he had eaten. 'I haven't been out of the cabin,' he said, 'since the night after the Titanic went down.'

An ambulance which had arrived to take Bride to hospital waited while Marconi heard the story of the disaster. On 19 April the *New York Times* declared in an editorial: 'If Guglielmo Marconi were not one of the most modest of men, as well as of great men, we would have heard something, possibly much, from him as to the emotions he must have felt when he went down to the Cunard wharf, Thursday night, and saw coming off the *Carpathia*, hundred after hundred, the survivors of the *Titanic*, every one of whom owed life itself to his knowledge as a scientist and his genius as an inventor.'

However, while the Americans were full of praise for wireless telegraphy, they were deeply unhappy about the way it had developed in the United States. Wireless's role in the *Titanic* tragedy caught the public imagination, and American Marconi's share price rocketed – which only served to emphasise that wireless was being controlled by an essentially foreign company. And some American Marconi operators were accused of encouraging Bride and Cottam to hold back their story when they might have given details from the *Carpathia*. The *New York Herald* discovered three marconiograms sent by W.T. Sammis, chief engineer of the American Marconi Company, to the operators on board the *Carpathia*. The first read: 'Keep your mouth shut. Hold story. Big money for you.' The second: 'If you are wise, hold story. The Marconi Company will take care of you.' And the third implicated Marconi himself in what the *Herald* and many politicians regarded as an unjustifiable news 'blackout' to further the company's own ends: 'Stop. Say nothing. Hold your story for dollars in four figures. Mr. Marconi agreeing. Will meet you at dock.' Marconi denied any involvement, while Sammis admitted he had sent the messages, but not until the *Carpathia* was near to docking in New York. Bride and Cottam did sell their exclusive stories to the *New York Times* for $1000 and $750 respectively, a fortune for young men earning the equivalent in pounds sterling of $360 a year.

This did not stop the Americans from treating the Marconi operators as honoured guests. Harold Bride was regarded as a hero by most of the newspapers, just as Jack Binns of the *Republic* had been. Jack Phillips, who had been buried at sea, was commemorated by statues in both America and his native England. Wireless had saved the lives of over seven hundred passengers on the *Titanic*, but the panic that followed the first news that the liner had hit an iceberg, and the chaotic whispering gallery of the airwaves over the following three days, finally drove the US government to seek to take control of this new technology. The owners of the *Titanic* were given a rough ride in Senate hearings to establish the cause of the disaster in which more than 1500 people perished, and Marconi himself was asked to explain and justify the behaviour of his American operators and managers. But the real scapegoats were the American wireless hams, many of them teenage boys, who were held responsible for much of the confusion about what had really happened on the night of 14 April.

Since 1909 several Bills had been introduced in the US Senate to regulate the amateurs. Boys as young as twelve had successfully argued against some of the proposals. The American Marconi Company had championed them, arguing forcefully that the real problem was the amateurism of the Navy and rival companies that had not got to grips with tuning. The boys who had built their own equipment were also potential recruits for the Marconi Company as it expanded and took on more and more operators. But after the *Titanic*'s sinking, though no amateur was ever found to have sent a misleading message, the Marconi Company quietly bowed to political pressure, and no longer opposed the regulation of amateurs.

Within months of the disaster all US amateurs were required to pass an examination before they could take out one of the newly issued licences. They were tested on their knowledge of Continental Morse code, and had to show that they could take apart and reassemble a small station. A great many just kept their heads down and did not bother to apply for licences. Of those who did, the vast majority passed the tests easily, and were amused that their examiners were drawn from the US Navy, whose operators they regarded as a joke.

Wireless hams were confined to a relatively short wavelength – two hundred metres or below – in the belief that this would stop them sending messages long distances. It was Marconi who was largely responsible for the notion that only very long waves generated by powerful transmitters and huge aerials could be sent and received over hundreds or thousands of miles. Short waves like those generated by Heinrich Hertz in his laboratory, for reasons Marconi did not understand, appeared to travel only short distances. But this was a fundamental error which arose from a failure to understand how wireless waves behaved, and was compounded by Marconi's early successes with long waves. Once it was understood that wireless signals of all wavelengths were reflected back to the earth from a conducting layer in the upper atmosphere, the reliance on long waves for distance was abandoned. Marconi acknowledged his error, which he discovered during the First World War while experimenting with very short waves. But in 1912 the potential of short waves for long-distance wireless communication was unknown, and the US government believed that restricting amateurs to waves of two hundred metres or less would dampen their enthusiasm, clearing the airwaves for the serious business of commercial and naval telegraphy.

The amateurs soon found their way round the regulations, but the trauma of the *Titanic* was a turning point in the early history of wireless telegraphy. Marconi was at the very height of his fame, and his companies were dominating the airwaves, but the spectre of government regulation in America was looming. The exciting pioneer days were over, and though Marconi and his engineers were to continue developing and improving the technology for many years, they would not remain leaders in the field for long. This was not because they failed to notice what was going on in America, where the spark transmitter was being replaced by high-speed alternators which could transmit speech, and improved versions of the de Forest audion were being adopted as the favoured receivers. Up to 1912 the company policy of sticking with well-tried technology had paid off, as the *Titanic* rescue dramatically demonstrated, and a vigorous enforcement of Marconi's patents had silenced the opposition. But the company had no claims on the new wireless technology, and could keep pace with it only by buying out the rights to it. It had the money and the will to do so, but history was no longer on Marconi's side – and by the autumn of 1912 it was apparent that his luck was running out.

41

✤➤➤⥼⥼⥼

The Crash

After the excitement of the *Titanic* rescue, Marconi went to inspect wireless stations in Pisa and Coltano. He took Bea with him, on the advice of family and friends who were anxious that the marriage was yet again under threat. As always, Marconi was fêted in Italy. King Victor Emmanuel and Queen Elena visited Coltano and invited them to their country home at San Rossore, near Pisa. Bea had brushed up on her Italian, only to discover to her dismay that the Italian royals spoke French most of the time; but she was honoured with an appointment as one of Queen Elena's ladies-in-waiting.

On 25 September the Marconis set off for Genoa, from where Guglielmo was to sail back to America, in a brand-new Fiat with Guglielmo at the wheel, Bea in the front passenger seat and the chauffeur and a secretary in the back. On a sharp bend they smashed head-on into another vehicle, and the Fiat was wrecked. The company magazine the *Marconiograph* reported: 'Mr Marconi was on the way to America when the accident happened at Foce, near Spezzia. His automobile was going at a moderate speed, but the other that ran into his was going very fast. One cause of these motor-car accidents in Italy is that the drivers are very careless about keeping the rule of the road. They run all over the place. Indeed both coachmen and chauffeurs act as if no rule of the road existed.' Everyone except Marconi himself escaped with minor injuries. The other vehicle had ridden up and over the Fiat, and in the impact Marconi had suffered a blow to the right side of his head. A car belonging to a local naval officer took him to hospital for first aid, and at first it was thought he had had a lucky escape. But the doctors discovered that his right eye was badly damaged, and that they could not save it. To protect the good left eye from sympathetic damage the right eye was removed at a clinic in Turin.

For a few days Marconi had a temporary glass eye. But he wanted the very best, and travelled to Venice to be fitted by the ageing Professor Luigi Rubbi, who was considered to be the most skilful maker and setter of glass eyes, reproducing not only the colour but the tiny veins and streaks of the original. It took nearly a week for the eye to be adjusted once Professor Rubbi had created it. Marconi's new eye was removable, and was taken out every night to be cleaned, like false teeth. Marconi's daughter Degna wrote later that it was a long time before she realised her father had a glass eye, which was a testimony both to Professor Rubbi's skill and to the very little time Marconi spent with his children.

Marconi's public adulation in Italy never dimmed. While he was recuperating in Venice in November 1912 he and Bea went to a performance at the Teatro Rossini opera house. According to the *Marconiograph*: 'As soon as his presence was observed the whole of those present rose and cheered, the ladies waving their handkerchiefs. Three times, Mr Marconi, who was greatly touched by the ovation, rose and bowed to the assembly.'

It was around this time that Marconi appears to have allowed his youthful caution and reserve to drop, and to make some uncharacteristically wild predictions about the future of wireless telegraphy. Interviewed by *Technical World* magazine, he was reported as saying:

> Within the next two generations we shall have not only wireless telegraphy and telephony, but also wireless transmission of all power for individual and corporate use, wireless heating and light, and wireless fertilising of fields.

When all that has been accomplished – as it surely will be – mankind will be free from many of the burdens imposed by present economic conditions. In the wireless era the government will necessarily be the owner of all the great sources of power. This will naturally bring railways, telegraph and telephone lines, great ocean-going vessels, and great mills and factories into public ownership. It will sweep away the present enormous corporations and will bring about a semi-socialistic state. I am not personally a socialist; I have small faith in any political propaganda; but I do believe that the progress of invention will create a state which will realise most of the present dreams of the socialists. The coming of the wireless era will make war impossible, because it will make war ridiculous. The inventor is the greatest revolutionist in the world.

The interview, published in October 1912, marks a turning point in Marconi's thinking, from pragmatist to visionary. From this time on he became more and more embroiled in politics – a subject he would have done much better to leave alone.

When he returned to London the following year, the newspapers were bristling with the story of the 'Marconi Scandal'. Before the *Titanic* disaster Godfrey Isaacs and Marconi had been able to raise capital for the American company by the successful issue of a new block of shares which were floated on the New York and London stock exchanges or sold to major US companies. Marconi took ten thousand himself, and Isaacs had one hundred thousand, a proportion of which he sold, some to members of his large family – he was one of nine children. Over lunch at the Savoy with his brothers Harry, who was in the fruit trade, and Rufus, who had risen rapidly as a lawyer to become Attorney-General in Asquith's Liberal Cabinet, Godfrey offered them American Marconi shares at a very favourable price. Harry took fifty thousand and then another six thousand to distribute amongst family members. Rufus took none immediately, but Harry soon persuaded him to buy ten thousand. Rufus then passed on a thousand shares each to Lloyd George, the Chancellor of the Exchequer, and Lord Murray, the Liberal Chief Whip.

When the Unionist and Conservative opposition got wind of this there were accusations of corruption. At the same time as the Isaacs brothers were dealing out American Marconi shares, the British Marconi Company was close to winning a contract from the government for the creation of what was known as the Imperial Wireless Scheme. The whole of the British Empire was to be linked by a series of wireless stations built and operated by the Marconi Company, which would profit enormously. In July 1912 a contract had been signed, but it had to be approved by Parliament. Was Rufus Isaacs buttering up the government with some potentially lucrative Marconi shares? Suspicions of skulduggery were heightened when Lloyd George and Lord Murray denied that they had bought any Marconi shares, but later admitted that they had been given shares in the American subsidiary. Investigations dragged on into 1913, and Marconi felt that his name was being dragged 'through the mire', although he was not in any way implicated himself. Though it was generally agreed that those members of the government involved had behaved unwisely, a Parliamentary Select Committee loaded with Liberals cleared them of misconduct, and the Imperial Wireless Scheme was begun towards the end of 1913.

Marconi, accustomed as he was to adulation, was shaken by the scandal. What he badly needed was another heroic wireless rescue at sea. The danger posed to shipping by icebergs was receding, as much more effective use was being made of wireless warnings. But in October a drama unfolded which was to place Marconi firmly back on his pedestal as the guardian of those at sea. This time the enemy which struck in mid-Atlantic was not ice, but fire. The saviour was once again what newspapers still called 'the magic of wireless'.

The *Volturno* was built in Glasgow in 1906 for the British Uranium Steamship Company, and named after the principal river in southern Italy. Its main function was to carry emigrants and cargo, and when it left the Dutch port of Rotterdam on 2 October 1913 for a regular run to New York with a one-day stop-over in Halifax, Nova Scotia there were only twenty-two first-class passengers out of more than five hundred on board. Most of those crammed below decks in steerage were poor Russians, Austrians, Croatians and others from Eastern Europe. Stowed beneath them were most of their worldly belongings, with which they hoped to start a new life in America. As well as its human cargo, the *Volturno* carried in its hold wines and spirits, barrels of tar and various chemicals. The crew of ninety-six were mostly Dutch, under the command of a British captain, Francis Inch. Manning the Marconi wireless cabin were twenty-two-year-old John Pennington and Walter Sedden, one year younger.

When the *Volturno* was in mid-Atlantic a storm blew up from the north-west. The steerage passengers had a miserable time huddled in their crowded quarters. A pull on a rough cigar, a pipe or a cigarette was some comfort, but it was strictly forbidden below decks, and carried a $5 fine. However, rules were always being broken, and if a steward came by the easiest way to avoid a penalty was to knock out a pipe or bury a cigarette between the deck boards. That was perhaps how the luggage stacked below steerage caught fire. In the early morning of 9 October smoke began to billow up from the hold as the *Volturno* rode the moun-tainous waves. The fire spread rapidly, and before it could be brought under control the liner was shaken by several explosions as the flames burned through to the tar barrels and liquor in the cargo hold.

While Captain Inch and his officers went to investigate, passen-gers on the upper deck, convinced that the ship would soon sink, launched two of the *Volturno*'s lifeboats. All those who scrambled and jumped into them were swept away, some under the ship's huge propellers, others smashed against the side of the liner by the heavy seas. They all died, adding to the toll of sixty crew and passengers killed in the explosions. Captain Inch then took control, and passengers were shepherded to the aft of the ship, away from the fire. Their only hope was that other ships riding the same Atlantic storm might come to their rescue. An 'SOS' was sent out from the Marconi cabin, giving the *Volturno*'s position, 49.12N 34.51W, and the message: 'Nos. one and two holds blazing furiously. Please come at once.' At just after 10 a.m. Sedden and Pennington received a reply from a fellow Marconi operator, P.B. Maltby, who was on duty on the Cunard liner *Carmania*, en route from New York to Liverpool: they were coming as fast as they could.

Captain Barr of the *Carmania* calculated that the *Volturno* was seventy-eight miles away. He ordered the stokers and engineers to bring the liner up to her maximum speed of twenty knots, and they rode into the storm. At the same time other ships heard the *Volturno*'s distress call, and turned about to go to her assistance. The *Carmania* relayed the 'SOS', and kept in constant touch with the *Volturno*. Four hours later they saw the blazing ship drifting helplessly in the storm. Captain Barr drew alongside and tried to throw out lines to the *Volturno*, but the sea was too rough, and he had to abandon the attempt for the safety of his own ship. One of the *Carmania*'s boats was lowered, but the crew gave up their rescue attempt after two hours. As night fell more ships arrived on the scene, but none could get close to the *Volturno* as the fire raged in the storm and spread to the midships. By nightfall nine ships, Russian, German, French and British, were standing by, able only to look on and wait for the storm to abate.

Captain Barr had a wireless message sent appealing for any oil tankers within range to join the rescue so that they might calm the seas by discharging some of their cargo onto them. A tanker of the Anglo-American Oil Company, the *Narragansett*, Morsed back: 'Yes, we will come with the milk in the morning.' The vigil went on through the night with the passengers and crews of the rescue ships watching in awe as the *Volturno* blazed, its terrified passengers screaming for help. From the wireless cabin came desperate appeals: 'For God's sake help us, or we perish.' But there was nothing that could be done until daybreak. Captain Barr asked the *Narragansett* if it could not make better speed. The tanker did its best, cutting an hour off its estimated time.

At dawn on 10 October the tanker appeared, and was positioned windward of the *Volturno*, from where it discharged two streams of lubricating oil. This quietened the sea enough for small boats to go out from the waiting ships to the *Volturno*. In the swell it was still difficult to manoeuvre beneath the burning liner, but by nine o'clock in the morning each of the rescue ships had taken on a share of the survivors. The German *Grosser Kurfürst* saved more than any other: sixty-seven passengers and nineteen crew members, who were taken on to New York. Another ninety survivors reached New York on board the Red Star Line's *Kroonland*. The *Carmania* took only eleven people, as it was too large to manoeuvre close to the *Volturno*. Seven other boats, including the tanker *Narragansett*, took those they rescued to England, France and in one case a return trip to Rotterdam. Parents and children had become separated in the confusion, and it was a long while before the fate of many passengers was known.

The rescue of the *Titanic* passengers had been a sensational success for Marconi's wireless. Because it was the largest liner in the world and had so many famous people aboard, the event received massive publicity. But the saving of those on board the *Volturno* demonstrated to shipping lines and the world's navies the huge potential of wireless for manoeuvring ships at sea. The whole operation, from the first messages to the *Carmania* to Captain Barr's call for an oil tanker to still the waters, had been entirely dependent on an agreed system of Morse code and distress signals, and free communication between ships of different nationalities. The wireless operators on the *Volturno* had been able to go on sending and receiving with emergency equipment after the liner's power failed.

On 15 October the London *Daily Telegraph*, comparing the triumph of the *Volturno* rescue with the tawdry matter of the Marconi shares affair, told its readers: 'the country on which he has showered such untold benefits has been content to single him out as an unwilling participant in an unsavoury scandal. Surely the time and occasion have arrived when the State may well revive, if that be necessary, its standard of honour, and grant to the wizard who enabled such a triumph to be achieved in the name of humanity some fitting token of England's gratitude for the great permanent addition he has made to what may be described as our armoury of mercy.' Even the satirical magazine *Punch* put aside its accustomed cynicism and published an illustration of a smart-looking Marconi at a ship's wireless station with the caption 'S.O.S', and Mr Punch himself, hat in hand, saying to the inventor of wireless: 'Many hearts bless you to-day, Sir. The World's Debt to you grows fast.' To add to his Nobel Prize and a host of other honours, Marconi received from King George V an honorary GCVO – the Victorian Order – as a personal gift. It was presented to him at Buckingham Palace in July 1914. A month later he was a suspect foreigner on British soil.

42

The Suspect Italian

On 5 August 1914, the day after Britain declared war on Germany, a British cable-laying ship, the *Telconia*, slipped quietly into the North Sea on a long-planned expedition. From the seabed it hooked up one by one five encrusted telegraph cables, then cut through them, making sure that if need be they could be reconnected to each other. These cables were the veins of Germany's Imperial communications network, running from the German port of Emden to Vigo in Spain, Tenerife in the Canary Islands, Brest on France's north-west coast, and on to America and the Azores. With these five cables out of action, Germany would have to rely heavily on wireless telegraphy throughout the four years of the First World War. At Nauen, just outside Berlin, the Germans had built the most powerful wireless station in the world, which could communicate with stations as far away as America and Togo in Africa. On 3 August an urgent message had been sent from there telling all German ships at sea to make for neutral ports. From the outset, wireless telegraphy changed the tactics of warfare, and the demands of war had a profound effect on the way in which the technology was exploited.

As has been noted, Germany, with few of its submarine cables remaining intact, became reliant upon wireless, while the British authorities were not at all sure what to do with it. Under the Defence of the Realm Act, hastily passed in August 1914, Britain's first move was to root out and close down anything that might be of use to foreign spies, and the newly appointed Secretary of State for War, Earl Kitchener of Khartoum, had no doubt where the greatest danger lay.

Shortly after war was declared, British police stations were issued with a set of photographs to enable them to distinguish instantly between potentially treacherous inhabitants and those who could be regarded as harmless. There were the 'carrier', the 'dragoon', the 'show homer' and the 'racer'. Armed with these mugshots, police constables set out to do their duty. Their investigations took them into the terraced streets of small towns in the Midlands and Lancashire, up back stairs, into secret lofts where the suspects were known to be lurking. Often when they arrived their quarry had flown, and they had to wait until the rattle of a trapdoor in the loft announced their return. The warning of Lord Kitchener was taken seriously: he did not want to see any homing pigeons flying around, for he feared there was a great danger of them being used to carry messages across the Channel. Suspected spies were tracked down: a German 'posing as a Dane' had his pigeon loft in Doughty Street, London, raided.

The talents of the tame blue rock dove, or homing pigeon, had been known for thousands of years. With careful training these birds could be released hundreds of miles from the loft where they were raised and fly non-stop at speeds of between thirty and sixty miles an hour back to the little trapdoor of their home. How they achieved this was no better understood than long-distance wireless telegraphy, though both were thought to be influenced in some way by the earth's magnetic field. In time of war, theory was of little value; but both wireless and winged messengers were to be of vital importance. Kitchener and the military authorities were persuaded that, if properly supervised, the homing pigeon fraternity could play a useful role, and that it would be better to license them than to ban them. A Voluntary Pigeon War Committee was formed, with the Marconi Company's managing director Godfrey Isaacs as the wireless telegraphy representative. Over half a million licences for the possession, carrying or releasing of pigeons were issued during the war years, including one to King George V, who had a royal pigeon loft on the Sandringham estate in Norfolk.

At the outset of the war the British Army imagined it could rely for its communications on the field telephone, telegraph cables and a few of the new-fangled wireless sets. But the highly efficient use of mobile pigeon lofts by the Germans and the French revealed the value of winged messengers when all else failed – as it often did when telegraph lines were blown to pieces by artillery bombardment. The British made great use of carrier pigeons during the Battle of the Somme, with four hundred messages flown back from the front on one September day, and five thousand sorties in all. The man who organised the operation, Lieutenant Colonel A.H. Osman, wrote in his account of the service *Pigeons in the Great War*, published in 1929: 'The advent and improvement of wireless has been the means of doing away with the use of pigeons for many services, but for espionage, secret service and many important duties, pigeons will never be replaced ... A pigeon silently flies through the air, there is no wave that indicates its use, nothing that indicates its point of departure or destination.'

Wireless amateurs were included along with pigeons in the Defence of the Realm Act: 'No person without the permission in writing of the Postmaster General shall buy, sell or have in his possession any apparatus for sending or receiving messages by wireless telegraphy, nor any apparatus intended to be used as a component part of such apparatus.'

Although the British Navy had been using wireless telegraphy in its manoeuvres and exercises for more than a decade, and a beginning had been made on constructing the chain of Imperial stations by the Marconi Company, few in positions of power yet understood the potential of this relatively new technology. Their first impulse was to blow up as many of the German wireless stations which had been established around the world as soon as possible.

Under the Defence of the Realm Act the Marconi Company

was taken over by the government in August 1914. The police also tracked down and closed 2500 licensed amateur wireless installations, and 750 that had been operating illegally. However, one or two stations avoided discovery, and continued to tune in after the declaration of war. Their defiance of the law proved to be of immense value, for they found they were picking up coded signals transmitted on predictable wavelengths by the German High Seas fleet, which was equipped with the very latest wireless technology.

Two of these amateurs – a barrister, Russell Clarke, and a retired colonel, Richard Hippisley – were very well connected. They were able to pass on messages they had received illegally to Sir Alfred Ewing, who had the title Director of Naval Education. Although eavesdropping on encrypted German wireless messages had been done before the war, the fact that potentially vital information could be picked up by stations in the West End of London suggested that some of the amateurs who had had their aerials taken down, and in certain cases their equipment confiscated or sealed in cupboards, might be every bit as useful as the patriotic pigeon-fanciers.

On 14 November the *Daily Telegraph* weighed up the pros and cons of wireless 'spying' in wartime:

> Wireless telegraphy conferred a boon upon mankind, but it is not without its dangers in times of international complications. A representative of the *Daily Telegraph* was yesterday shown messages originating in Germany, France and the North Sea, which some time ago were received at a private wireless station in the West End. Like the telephone when it was in its infancy, the wireless system attracted many amateurs and experimentalists, and numbers of aerials were erected. In time of war these installations may be used against the public weal. They may also be brought to serve in the best interests of the Empire by 'catching' stray messages intended for the enemy.

Germany had created an ingenious range of diplomatic and military codes, and was confident that any messages picked up in

Britain or America would be unintelligible to their enemies. The German High Seas fleet made free use of wireless when it was on the move, sending Morse messages in code to Zeppelins and U-boats. Great advances had been made in transmitter technology, and since around 1905 the old intermittent spark transmitters had been replaced by rotary spark generators which sent a continuous signal that was musical in tone (it was said to 'sing') and gave wonderfully clear dots and dashes. If the British could eavesdrop on these messages and crack the codes, Germany's war plans would be an open book. A large number of the Marconi Company's skilled operators who were able to accurately record high-speed Morse code volunteered for active service, and while more young men were being trained by the Naval Education Department, the only available pool of expertise was in the Wireless Societies that had been formed by amateurs.

The two amateurs Russell Clarke and Richard Hippisley, who had alerted the Admiralty to the possibility of tapping German wireless messages, were sent to Hunstanton on the north Norfolk coast to set up a listening station, and the government surreptitiously began to bring into the war effort the wireless enthusiasts whose equipment it had previously wanted to silence. They were not to transmit anything: their job was to listen in. Over a period of time these vigilantes were sworn in as special constables, and took a pledge in which they 'solemnly and sincerely' declared that they would not reveal to any 'improper' authority the contents of any messages received. Although nobody thought of them in that way at the time, this amateur wireless fraternity was the first 'audience' for broadcasts in Britain. They thrilled not to the sound of music, or reports of great sporting events, but to messages sent as series of numerals which when deciphered could play a crucial part in the outcome of the war.

The first network of listening stations was established along the eastern coast of Britain to monitor the movements of the German High Seas fleet, which would make sorties into the North Sea with the intention of picking off patrolling British Navy vessels. The hope of the British Admiralty was that the navy could catch the main part of the German fleet in open sea, and inflict a devastating defeat upon it with its superior firepower. This cat-and-mouse strategy was played out numerous times throughout the first year of the war.

Wireless messages picked up by the vigilantes were sent to a team of code-breakers who worked in great secrecy in Room 40 of Admiralty House in Whitehall. When they cracked a code and were confident that they knew what the German fleet was up to, the information was passed on to the navy. In theory it was like a game of chess in which one side knows what the other's next move is going to be. In practice, however, a great deal of valuable information was misused or wasted because of a distrust among naval command of the accuracy and value of what they were being told.

Among the many Marconi Company technicians who volunteered for active service was the engineer H.J. Round, a favourite of Marconi himself, who before the war had been developing the 'valve' receiver that Sir Ambrose Fleming had devised in a moment of inspiration in 1903. The valve receiver, which had been used experimentally for some time, had tremendous advantages over the crystal set or the crude but reliable 'Maggie': it was more sensitive, could operate at higher Morse speeds and could also carry speech. At first Round was sent to the Western Front, where he discovered that diode valve receivers could be used for pinpointing where a wireless signal had come from. He returned to Britain and set up direction-finding stations on the east coast which could plot the path of Zeppelins as they acted as lookouts for the German fleet or set out on bombing raids over England. In this way the nature of warfare was transformed by wireless, which proved of great value to Britain not so much as a technological advance on pigeons or the traditional semaphore of naval communications, but for sophisticated espionage. It was the beginning of the secret tapping of enemy signals which would play such a vital part in British intelligence during the Second World War.

The position Guglielmo Marconi found himself in as an Italian national in Britain was difficult. At the outset of the war Italy had remained neutral, but there was reason to believe that it would soon enter the conflict as an ally of Germany. Bea, Degna and Giulio were at Eaglehurst in August 1914, and keenly felt the local suspicions that Signor Marconi's wireless station in one of the old towers might be used for espionage. Degna recalled her mother being told she was not allowed to leave the grounds of Eaglehurst for several weeks, before the anxiety died down. When Marconi applied to the Home Office for exemption under the Aliens Restriction Order, so as to be able to travel around the country, he was at first refused. In time, however, his impeccable social and political connections freed him from suspicion, and the ban was lifted. In the winter of 1914 he crossed the Channel and took the train to Italy, which still maintained its neutrality.

Bea's sister Lilah and Marconi's Italian secretary, a Mr Villarosa, had been stranded in Rome at the outbreak of war, but Marconi was able to arrange for them to travel to England with the protection of the British diplomatic courier. He stayed on in Rome, where he had been given a seat in the Senate, and began what was to be a chequered political career. Italy had still not declared its hand early in 1915, and at the end of April Marconi took the Cunard liner *Lusitania* to America, where the American Marconi Company was still operating, running the gauntlet of the German U-boats which had begun to attack ships in British territorial waters. While Marconi was giving evidence in an American court in yet another patent dispute, the *Lusitania* set out on its return voyage to Liverpool on 1 May, with 1257 passengers and a crew of more than seven hundred. All had been warned by a notice posted by the Cunard Line:

> Travellers intending to embark on the Atlantic voyage are reminded that a state of war exists between Germany and her allies and Great Britain and her allies; that the zone of war includes the waters adjacent to the British Isles; that, in accordance with formal notice given by the

Imperial German government, vessels flying the flag of
Great Britain, or any of her allies, are liable to destruction in those waters and that travellers sailing in the
war zone on ships of Great Britain or her allies do so at
their own risk. IMPERIAL GERMAN EMBASSY
WASHINGTON, D.C., APRIL

As more than a hundred of the passengers on the *Lusitania* were
Americans, and it was assumed that Germany would not want to
bring such a powerful ally of Britain into the war, there was no
great fear that the vessel was in serious danger of attack. What the
passengers did not know was that the *Lusitania* was carrying
a secret cache of munitions for the British. It is possible that
the German Admiralty had been passed this information from
the powerful Sayville wireless station on Long Island, which was
owned and run by a subsidiary of Telefunken, named the Atlantic
Communications Company. The United States allowed wireless
messages to be sent across the Atlantic provided they were 'neutral'
– that is to say, not concerned with the war. Plain-language messages only were allowed, and American censors were posted to
both American Marconi stations and to two large German-owned
stations at Tuckerton, New Jersey and Sayville. But, just two days
after Britain declared war on Germany, the *New York Times* had
carried a story claiming that the Sayville station was breaking the
rules and passing valuable information on shipping to the German
navy, which had begun to patrol the western coast of Britain with
U-boats.

On 7 May the *Lusitania* reached the Irish Channel, and was
making slow progress towards Liverpool in heavy fog. The liner
was spotted by Captain Walther Schwieger of the submarine
U-Boat 20, which had been harassing shipping for several days.
He regarded the *Lusitania* as a legitimate target, and at 1.20 p.m.
he came to the surface and fired two torpedoes at close range.
Lookouts saw the danger too late for the *Lusitania* to alter course
or pick up speed. There were two explosions, and the vessel listed

heavily, so that it was difficult to launch the lifeboats, which were either hanging from the raised section of the hull or buried beneath the rapidly submerging section. At 2 p.m. the *Lusitania* sank in shallow water, with its bow still showing above the waves as it rested on the seabed. Among the 1198 men, women and children who died were 124 American citizens. There were only 761 survivors.

Two weeks after the sinking of the *Lusitania* had outraged the United States – without shaking its determination to remain a bystander in the European conflict – Marconi boarded the American liner *St Paul* to return to England. He had been told by the Italian Ambassador to Washington that an agreement with Britain and France meant that Italy would soon be at war, and had asked the American judge hearing his patent case to release him. When Italy declared war on Austria-Hungary on 24 May, Marconi was well out into the Atlantic. It was rumoured that there was a German plot to capture him, and that news of his presence aboard the *St Paul* had been tapped out in code by the treacherous Sayville transmitter and picked up by the powerful Nauen station in Germany. Marconi, it seemed, was in danger of being tracked down by his own invention. In fact his departure on the *St Paul* had been front-page news in the *New York Tribune* on Sunday, 23 May. He was pictured with a lady war correspondent headed for the front. This was none other than his ex-fiancée, who was now married: Mrs Inez Milholland Boissevain. No mention was made of their former association.

After a short visit to London and Eaglehurst, Marconi travelled on to Paris and then Italy, where in June he was commissioned as a lieutenant in the Italian army, with responsibility for overseeing wireless equipment. He stayed in Italy for the rest of the war, developing directional short-wave wireless for the Italian navy. Bea and the children joined him, and in 1916 their third child, Gioia Jolanda, was born in Rome. Meanwhile, America finally stole Marconi's thunder.

43

✥ ✥ ✥ ✥ ✥

Eclipse of Marconi on the Eiffel Tower

When in January 1887 the engineer Gustave Eiffel and his team won the prize to build a tower for the Paris World Fair of 1889, there were protests from some of the most prominent writers and intellectuals of the day. A letter signed by Alexandre Dumas, Guy de Maupassant, Paul Verlaine and others began: 'We, the writers, painters, sculptors, architects and lovers of the beauty of Paris, do protest with all our vigour and all our indignation, in the name of French taste and endangered French art and history, against the useless and monstrous Eiffel Tower . . .'

At the time it was completed in the spring of 1889, Eiffel's tower did appear to be nothing more than an engineering folly, comprising more than eight thousand individually crafted metal sections held together by over eight million rivets. But as an attraction it was a spectacular success: nearly two million people paid to climb it or take the lifts to the top between May and November 1889, and by December admission receipts had recouped three quarters of the eight million gold francs it had cost to build. It was lit at night, first by twenty-two thousand gas burners, a few years later by electricity. Very soon the 'monster' became France's most distinctive and cherished symbol, and it was not long before a use was found for it. Towering a thousand feet above the rooftops of Paris, it looked for all the world like a gigantic Marconi radio mast; and when experiments with wireless began, that is what it soon became.

The Frenchman Eugène Ducretet made the first transmission from the tower's summit to the Pantheon four kilometres away in 1898, and seven years later another inventor, Captain Gustave Ferrie, established a wireless telegraphy link with French forces on the German borders. In 1906 Ferrie was communicating with French ships at sea, by 1907 he was able to exchange messages with Casablanca in Morocco, and by 1908 his signals from the Eiffel Tower reached a distance of four thousand kilometres.

When Germany invaded Belgium in 1914 with plans to occupy France, its great prize was to be Paris. The Eiffel Tower became the centre of French military wireless communications, and there were plans to destroy it if the Germans succeeded in breaking through to the capital. But the invaders were halted sixty miles from Paris, and had to dig in along what became known as the Western Front. The Eiffel Tower remained secure, supporting one of the most powerful wireless stations in the world throughout the war. And in the midst of the conflict it was to play a part in a revolutionary breakthrough in wireless which made headline news on both sides of the Atlantic. Neither the name of Marconi nor that of anyone associated with his companies, which were working flat out to equip the British Army and Navy with wireless telegraphy transmitters and receivers, appeared in the sensational stories of the first-ever transmission of the spoken word, from three thousand miles away across the Atlantic to the top of the Eiffel Tower.

The language of the first wireless telephone message was English, the accent American. An engineer speaking into a microphone at the US naval wireless station established a few years earlier at Arlington, Virginia, was heard by a fellow American high up on the Eiffel Tower. It was only possible to send west to east, and confirmation that the engineer's voice had been heard distinctly was cabled back across the Atlantic. The French, naturally preoccupied with their military communications, allowed the Americans only a few minutes over a period of several days to make this historic broadcast. As with Marconi's first signal across the Atlantic, it was still reliant on cables. Telegraph messages were sent to alert

Paris to the times when attempts at speech transmission would be made. But there was no longer a conflict between cable telegraphy and wireless, for most of the technology was provided by the giant US company American Telephone and Telegraph – AT&T. The announcement of the successful transmission of speech was made by the company's dynamic head, Theodore N. Vail, on 22 October 1915.

At the time Europe went to war, the evolution of the 'valve' first devised by Ambrose Fleming had been very rapid. Lee de Forest had turned it into the audion, and had discovered that it could be used for amplifying signals received. A young New York amateur enthusiast, Edwin Armstrong, had improved on de Forest's audion, and discovered that it could be used not only for wireless reception, but for transmission as well. Inevitably Armstrong and de Forest became locked in a series of patent disputes, but AT&T had the power and the money to buy up the latest and best technology and to develop it themselves.

Theodore Vail wanted to stay in the forefront of wireless telephony developments so that his company's already extensive cable telephone network would not have any rivals. The transmission of speech would mean that ship and shore could speak to each other without the need for specialist operators who could read Morse code. Out-of-the-way places could be reached by telephone without cables. In October 1915, after AT&T had shown that the wireless telephone could span the North American continent, and just before the announcement of the transatlantic triumph, Vail told the *New York Times*: 'What impresses me is that we, in America, are doing this work, this big, constructive work, at a time when all Europe is at war. That's pretty fine.' Lee de Forest did not agree. He believed it was his audion which had provided the breakthrough, and tried to convince the press and the public that Vail had stolen his invention. He even travelled to Paris at his own expense when the Eiffel Tower transmission was made, but was barred from any involvement.

The United States had the field to itself for two and a half vital years in the development of wireless. AT&T was investing heavily in wireless telephony, and the US Navy, which had dillied and dallied over which system to buy, now equipped itself with the very latest technology. The large and growing fraternity of American amateurs were now forming themselves into more effective political groupings, and had become far and away the world's largest civilian 'audience' for Morse messages sent by wireless. Despite the regulations which had followed the *Titanic* disaster, American amateurs had a fairly free hand, and were keen to buy or make the very latest devices available. While their enthusiasm was directed towards the peacetime uses of wireless, Europe was adapting an already anachronistic system of wireless telegraphy to warfare. Marconi was not out of the picture, but his pre-eminence in the field of wireless was on the wane. When the war was over Guglielmo Marconi could at last rest on his laurels, and enjoy an entirely new era of radio broadcasting to which his boyhood invention had given rise, but in which he himself played only a minor part.

44

❧❧❧❦❦❦

In Bed with Mussolini

The Villa Griffone, where Guglielmo Marconi spent so many of his youthful hours experimenting with electricity, is in many ways little changed. In the summer heat the stony earth shimmers, and in the evenings the invisible cicadas fill the air with their incessant chattering. At night the little Scops owls call with a single note which has an electronic sound, a 'pink, pink', like a lazy Morse 'dot, dot, dot'. The vineyards are still tended and the lawns around the villa watered. Everything is charming – until you stroll below the villa, down the hillside that falls away into a river valley where Marconi played as a boy. Into this hillside a kind of bleak, featureless amphitheatre has been cut, framing a grotesque pseudo-classical stone entrance to a subterranean vault. Marconi himself never saw this. It is a mausoleum, commissioned by Benito Mussolini to commemorate the great Italian inventor's death on 20 July 1937. Marconi's body was taken from a local cemetery to be interred in this monstrosity, and that is where he still lies, oblivious to the atrocities which were committed in World War II by his friend and admirer Mussolini and his Nazi allies. Recent allegations that Marconi secretly prevented Jews from joining the prestigious academy he headed from 1930 come as a shock, but are also a reminder that his dogged refusal to bow to scientific authorities was not matched by a similar independence in his political life. Marconi became close friends with Mussolini, and actively promoted the fascist cause in the last years of his life. The mausoleum is a fitting reflection of what Marconi's political beliefs would become during those years.

Elsewhere in the grounds there is a stone bust of Marconi facing away from the villa. On close inspection there is a pockmark made by a bullet in the back of the head, an idle act of vandalism perpetrated by a German soldier billeted at Griffone during the Second World War, for which an apology was later received. Towering over everything is a giant metal sculpture of Marconi, black and sombre, and close by an intriguing piece of mangled metal that a small plaque identifies as the only remaining relic of a beautiful white steam yacht which Marconi had owned and on which he spent a large part of the years before his death.

In contrast to the hideous, fascistic bunker in which Marconi is buried is the beautifully restored apparatus used in the inventor's youthful experiments in the airy rooms of the Villa Griffone itself. The worktable in his recreated attic laboratory, with a window that looks out from the back of the villa to the vineyard on the hillside, is strewn with all the enthusiastic amateur's make-do bits of equipment: glass vials, rolls of wire, archaic-looking batteries, metal filings and the little hand-bellows used for moulding the miniature coherer. It is hard to imagine now that this was the birthplace of what was regarded at the time of Marconi's death as the most spectacular and influential new technology of the twentieth century. Though he had long before lost the lead in the development of what became known as radio, Marconi's world fame had not yet waned in the 1930s, and when news of his death was transmitted around the world every newspaper devoted pages to the details of his state funeral and the story of his remarkable life.

By the time the armistice was signed in November 1918 wireless had changed almost beyond recognition. It would be only four years before the broadcasting of radio programmes swept across America like a smouldering fire that had suddenly burst into flame. Marconi's little boxes became museum pieces, as the valve replaced the spark and the crystal set and the 'Maggie'. The Marconi

Companies in the United States and Britain had both been taken over by government during the war, and there was a reluctance to allow them to regain their previous dominance. The US Navy wanted to control all wireless in America. It lost that battle, but the government engineered the creation in October 1919 of a giant new company, the Radio Corporation of America, RCA. In effect American Marconi was forced to sell up as RCA bought out every American patent relating to radio, including those of de Forest and Fessenden which had fallen into the hands of other companies.

It took several years for the British government to decide on the future of radio broadcasting. New licences for wireless 'telephony', the transmission of speech and music, were issued by the Post Office, and in 1920 engineers of the Marconi Company began to play records and concerts over the air from a small studio at their offices in Chelmsford, Essex. The very first advertised broadcast, which could be heard by only a few enthusiasts with crystal sets or valve receivers, was made from Chelmsford at 7.10 p.m. on 15 June 1920, with the sponsorship of the *Daily Mail*. 'The Australian nightingale' Dame Nellie Melba sang three songs into a telephone microphone with a cigar-box 'horn': 'Home Sweet Home', 'Nymphs and Shepherds' and the 'Addio' from *La Bohème*.

Broadcasting of this kind was a new departure for the Marconi Company, which soon began to manufacture radio receivers for the general public. In 1923 it became part of the British Broadcasting Company, making a few hours of programmes a week under strict control by government. To forestall a broadcasting free-for-all like that in America, in 1926 the government created the British Broadcasting Corporation, which had a monopoly on all programmes. The Marconi Company continued to thrive as a major supplier of crystal sets and valve receivers to the rapidly growing population of 'listeners', as well as to aircraft and the military. The Morse code era was not over: wireless telegraphy was used by ships for many years after broadcasting began, as a relatively cheap and efficient form of communication.

Through all these momentous changes, Marconi himself steered his own course, as single-mindedly as ever. Though his company had made the first radio broadcasts in Britain, he took little interest in this new form of entertainment, except as one of the growing number of listeners. In 1919 he began negotiations with the British Admiralty to buy a steam yacht that had been taken as a spoil of war. It had been built in Scotland and was owned originally by the Archduchess Maria Theresa of Austria, who had christened it the *Rovenska*. To help pay for it Marconi had sold the family home in Rome, very much against the wishes of Beatrice and the children. By February 1920 the yacht had been completely refurbished under the supervision of the ever loyal George Kemp, and renamed *Elettra* – electricity – and Marconi took his family and some guests on a cruise of the Mediterranean. The children loved the ship with its crew of thirty, its saloons and wireless cabin. But it was the end for Beatrice when she discovered that among those on board was Marconi's latest conquest. Bea had known that he had had lovers before, and had forgiven him. But now she was prepared to let him sail away on his own, and they finally separated.

The *Elettra* became Marconi's home and his laboratory. He was still obsessed with the distance wireless waves could be sent, and sailed the world's oceans experimenting with short waves transmitted from his yacht to shore stations. The inhibiting influence of daylight, which had always baffled him, continued to have inexplicable effects on all the different wavelengths he tried, though he was now sending and receiving signals over distances of more than four thousand miles. Marconi continued to mix in the most elevated society, welcoming kings and queens aboard the *Elettra* and enjoying discreet affairs with an unknown number of elegant ladies. When his mother died in London at the age of eighty and was buried in Highgate Cemetery, Marconi was in Italy, and did not get to the funeral.

In October 1922 the Italian Fascist leader Benito Mussolini came to power after the march of his supporters on Rome, and Senatore Marconi soon became a close friend and supporter. The following year Beatrice, who had found a lover, the Marchese Liborio

Marignoli, whom she wanted to marry, asked Marconi for a divorce. This was no simple matter in Catholic Italy, but by making themselves temporary citizens of the free city of Fiume the divorce went through in February 1924 on the grounds of Bea's adultery, and in April she married Marignoli.

A free man on his great white yacht, Marconi was rumoured several times to be about to remarry. He wrote to Beatrice during these years as if to a friend and confidante, saying on occasion how lonely he was. In 1926 he fell in love with a beautiful and much younger Italian woman, Cristina Bezzi-Scali, but because he was technically a Protestant and divorced, and she was from the Catholic nobility – her father was in the Vatican – marriage appeared to be impossible. As always, Marconi consulted the experts, and found that it might be possible for his first marriage to be annulled if it could be shown there was 'not proper consent' to it. Beatrice agreed to connive in this fiction, and the annulment was granted after Marconi had given evidence before a commission at Westminster Cathedral in London.

It was in that year, 1926, that the puzzle about how wireless waves travelled around the world was finally solved. Marconi had always believed that only very long waves could travel any distance, but during the war he had experimented with very short waves, which he concentrated into a beam with the idea of improving ship to ship communication. He discovered that these short waves could travel great distances; amateur radio enthusiasts in America, confined by law to short waves, made the same discovery. It was an experiment carried out by an Englishman, Professor Edward Appleton of Cambridge University, which provided the explanation. A directional beam was transmitted vertically upwards and it bounced back. Oliver Heaviside, the eccentric Hermit of Paignton who had died in obscurity in 1925, despite the efforts of his many admirers around the world to persuade him to accept the recognition he deserved, had got it right back in 1902. The ionosphere which reflected wireless signals turned out to be a complex three-layered sandwich of particles which had a variable effect on wireless waves of different lengths.* Its composition was affected by the sun's rays – which explained the difference in the distance signals would travel during the day and the night. This finally put paid to a theory of the writer Sir Arthur Conan Doyle, an enthusiastic spiritualist, that the greater distances achieved by Marconi at night were proof of the mysterious 'powers of darkness' which spiritualist mediums exploited.

Professor Appleton paid glowing tribute to Marconi in an appreciation published in the *Daily Mail* on 21 July 1937, the day after Marconi's death:

> The magnitude of the benefits conferred on humanity by Senator Marconi's radio discoveries needs little emphasis in a world which is daily making use of them. But what is not always realised is the fundamental importance of some of his great technical achievements, which have been the starting point of major advances in scientific knowledge. These achievements can all, I think, be traced to *Marconi's almost obstinate belief that there is no limit to which wireless waves will not travel* [Italics in original] ... His motto, like that of the great Faraday himself, was '*Try it*.'

The distinguished former rival of Marconi, Sir Oliver Lodge, had continued his pursuit of the spirits while Marconi was discovering the power of short waves, which cracked the problem of daytime transmission at a distance. If anything Lodge's belief in the afterlife had become stronger since the death of the youngest of his six sons, Raymond, in the trenches of the First World War (Lodge had six daughters as well, twelve children in all). The terrible news that Raymond had died of shrapnel wounds arrived by telegram in mid-September 1915, and within a few days Lodge and his wife were consulting mediums in the hope of receiving messages from their son. Lodge even believed that he had been

* It was found that ultra-short waves were reflected by a different layer of the atmosphere than long waves.

warned of Raymond's death by his old friend, the late Frederic Myers, in a 'cross correspondence' message. In 1916 Lodge published a sensational book, called simply *Raymond*, giving an account of his spiritualist experiences. At the time there was massive popular interest in the afterlife, as so many families had lost sons whose bodies they would never see, and who would have no proper funeral.

Paradoxically, it was Marconi's discoveries with wireless that allowed Lodge to test a spiritualist theory. In 1927 Lodge asked the first Director-General of the BBC, John Reith, if he could carry out an experiment in telepathy in a radio broadcast. It was organised by the Society for Psychical Research, whose members attempted to 'transmit' through the ether to listeners all over the world a set of images – a bunch of white lilac, a man in a mask, a Japanese print and two different playing cards – that they were shown. The hope was that listeners would pick up the 'transmitted' images telepathically. Twenty-five thousand impressions were sent by listeners to the SPR offices from around the world, but they bore little resemblance to the images that the SPR members had viewed, and the experiment was judged at best inconclusive, at worst a failure. Lodge continued to believe, however, that both spiritualist messages and wireless waves could travel through the 'ether', though by the 1930s scientists were beginning to realise that that mysterious substance had been no more than a figment of the imagination. Quietly, the ether theory drifted into history.

Marconi had no interest in spiritualism, and always became short-tempered when, as frequently happened, he was asked if he might soon be able to communicate with Mars. He had no time for such idle speculation, being more preoccupied with the complications of his private life. These were resolved when he married Cristina in April 1927. Three years later their only child was born, a girl who was named Maria Elettra Elena Anna. She was cared for much of the time by her grandmother Countess Bezzi-Scali in Rome, while Marconi and Cristina continued to travel the world, often on board the *Elettra*. Wherever they went Marconi was greeted as a distinguished guest. The honours accumulated on him like crustacea on a cruising whale. He continued to experiment, chiefly while at sea on his yacht, which was now able to pick up wireless broadcasts from all over the world.

In the last years of his life Marconi abandoned the Villa Griffone and lived in a sumptuous apartment in Rome. He had suffered several heart attacks, and became seriously ill when on a visit to London in 1934. Despite his failing health, and against the advice of his doctor, he took up the cause of Italy when Mussolini invaded Abyssinia the following year. He sailed to Brazil as the Fascist leader's envoy in October 1935 to seek the support of the many Italians living there, and then to England, where he met Edward VIII at the time of the abdication crisis. On his travels he had made many broadcasts putting the Italian case in the Abyssinian conflict, but when he sought permission from the BBC to do the same in Britain he was told politely that while he could broadcast about anything he liked, he would not be allowed to put Mussolini's case on British radio.

Marconi continued to support Mussolini, and was due to meet the Italian leader at six o'clock on the evening of 19 July 1937 to talk about his latest experiments with ultra-short-wave radio. In the morning he took Cristina and Elettra to the station: it was Elettra's seventh birthday the following day and they were going to Viareggio, expecting Marconi to follow. But in the afternoon he became too ill to leave his apartment. He had converted to Catholicism after his second marriage, and a priest was called to administer the last rites. When the news of Marconi's death broke early the following morning, the first person at the deathbed was Mussolini, who ordered that a state funeral should be held.

EPILOGUE

In 1894, when Guglielmo Marconi was twenty years old, there was no such thing as wireless; when he died in 1937, at the age of sixty-three, the first television broadcasts had been made, and a wireless signal had been sent around the world. Many of his obituaries compared him with Christopher Columbus: the explorer had dismissed the idea that the world was flat, and had the temerity to sail over the horizon; Marconi had ignored the theory that because the earth was round, wireless waves could not travel very far. By general agreement his most important technical achievement was the transatlantic transmission of the letter 'S' in Morse code from Poldhu to Newfoundland in December 1901, and the greatest value of his magic box – reference was still made to the miraculous nature of the invention – was the saving of lives at sea. Newspapers around the world recalled the triumph of the *Titanic* rescue.

Marconi's funeral rites were long-drawn-out. There was a procession from the Farnese Palace in Rome to the church of Santa Maria degli Angeli, where thousands filed past his open coffin as he lay in state. There was a further funeral ceremony in Bologna. The day after his death the *New York Times* devoted a whole page to his achievements, under a headline that ran across eight columns: 'World-Wide Tribute is Paid to Marconi as the Benefactor of Many Millions'. American radio stations broadcast programmes recalling his achievements, and much was made of the fact that it was his invention that made it possible for an audience across the continent to listen in. Lee de Forest, who had finally made some money when his disputed patents were bought out by AT&T, and who had awarded himself the title 'the Father of Radio', commended Marconi for his 'daring genius'. He could rightly be called 'the father of the wireless telegraph', de Forest said magnanimously of his old rival, regretting that in all the years their paths had never crossed.

On 21 July, while remembrance services were held in London for Marconi, the General Post Office decreed that from 6 p.m. there should be a two-minute radio silence for all but distress calls. At Broadcasting House in London, the headquarters of the BBC, the flags flew at half-mast. In the United States, too, radio silence was observed as a mark of respect, and a special tribute was paid by David Sarnoff, President of RCA. An immigrant from Russia as a child, Sarnoff had worked for the American Marconi Company before the First World War, and had been one of the operators at the station in New York which received news of the loss of the *Titanic* in 1912. In Italy Mussolini ordered five minutes' radio silence and the closing of shops and businesses.

Very little was said in the public tributes about Marconi's private life, and the remark was often repeated that he was 'wedded to wireless'. His first wife Beatrice joined the solemn crowds which had gone to see Marconi lying in state, and paid her last respects without anyone noticing her or knowing who she was. According to the newspapers Marconi had made a will in April 1935 which was just a page and a half long. His fortune reportedly amounted to five hundred million Italian lire, or £5 million* (some later estimates put the figure at £1.5 million), most of which was left to his youngest child, Elettra, with the proviso that she support her mother from the income from the estate. Much would go on death duties, but there was no doubt that Marconi had amassed a considerable fortune.

In the week following Marconi's death the story of how the young Italian came to London with his mother in 1896 was told over and over again in the newspapers. The *Daily Telegraph*'s 'Radio Correspondent' L. Marsland Gander recalled the help Marconi

---

* Roughly £150 million at current values.

had been given by William Preece of the Post Office. It was a 'lasting credit' to Britain that it had shown such faith in a pioneer scientific effort when all Marconi had to offer was 'a magic tin box on a pole'.

The speed with which Marconi developed his invention, and his sheer audacity, astounded his contemporaries. Preece gave him his first break, but failed to keep pace with his protégé. As Thomas Edison said: 'Marconi delivers more than he promises.' That was true of him during the most exciting, inventive years of his career, between 1896 and the outbreak of the First World War. But Marconi became a victim of his own success, neglecting his domestic life in favour of a single-minded and fanatical pursuit of invisible waves. The young man alone in the attic laboratory of the Villa Griffone ended up sailing the world's oceans on his beloved white yacht, in whose wireless cabin he could escape from the pressures of fame and fortune. There is a sense in which Guglielmo Marconi was somehow always alone, mesmerised by his own magic, the workings of which he never really understood.

INDEX

Admiralty (British): experiments with GM's system, 57; denies help to Marconi Company, 196; in World War I, 273

aerials *see* directional aerials

Alexandra, Queen of Edward VII, 134

Allan Line (shipping), 88

alternators, 259

American Institute of Electrical Engineers, 110–12

American Marconi Company, 43, 176, 240, 256–7, 263, 283

American Telephone & Telegraph (AT & T), 279–80

American Wireless Telephone and Telegraph Company, 116–19

America's Cup (yacht races), 57, 59–62, 83, 86, 117, 154

Ampthill, Emily, 42

Andrews, Thomas, 249

Anglo-American Cable Company: completes first transatlantic cable, 89; claims breach of monopoly

against GM's transatlantic signals, 102–4, 112

Appleton, Edward, 285–6

Arco, Georg, Count von, 75, 136–7, 139, 176, 185

Armstrong, Edwin, 279

Armstrong, William, 22

Athern, 'Pop', 163

Atlantic Communications Company, 275

Atlantic Ocean: steamships on, 64–5; GM aims to bridge with wireless signal, 65, 67, 74, 76–7, 81, 90, 107, 125, 141; signals first cross, 100–1, 110; cables, 105; signals sent between Poldhu and Glace Bay, 143–9; GM publishes first shipboard regular newspaper (on *Lucania*), 159; first spoken words transmitted across, 278

'audion' receiver, 207–8, 218, 259, 279

Aurania (ocean liner), 57–8, 65

Baden-Powell, Baden: kites, 46, 100

Baker, Ray Standard, 104

balloons *see* kites and balloons

Baltic, SS, 220–3

Barber, Commander Francis M., 176

Barnett, Canon Samuel Augustus, 1

Barnum, Phineas T., 168n

Barr, Captain (of *Carmania*), 265–6

barreter (receiving device), 84, 86, 117, 187

Bathurst, Sir Richard Harvey, 217

Belgian Marconi Wireless Company, 193

Bell, Alexander Graham: invents telephone, 6, 49, 95; demonstrates to Queen Victoria, 26; in New York audience honouring GM, 112; wife's deafness, 143n

Bennett, Gordon, Jr, 57, 60, 102, 180

Bennett, James Gordon, Sr, 60

Berlin Radiotelegraphic Conference (1906), 222

Binns, Jack, 220–4, 257

Bismarck, Prince Otto von, 136

Blok, Arthur, 142–3, 167–9

Boer War, 69, 160

Boissevain, Inez *see* Milholland, Inez

Bologna, 10–11, 157, 180

Booth, William, 1

Bose, Jagdish Chandra, 5

Bottomley, John, 240, 254–5

Bournemouth, 36–7

Bovill, W.B. Foster: *Hungary and the Hungarians*, 212

Boxall, Joseph, 251

Boyle, Sir Cavendish, 101

Bradfield, W.W., 69, 76, 235

Bradford, Admiral Royal Bird (USN), 176

Branly, Edouard, 17, 98

Brant Rock, Massachusetts, 157, 186, 208

Braun, Karl Ferdinand, 204, 228

Brazil, 288

Bride, Harold, 244–5, 247–8, 250–1, 253–7

British Broadcasting Company, 283

British Broadcasting Corporation (BBC), 283, 290

Brown, Harry, 163

Brownsea Island, Dorset, 178–9, 182–3

Bruce, SS, 104

Brunel, Isambard Kingdom, 89

Budapest: early telephone broadcasting system, 211–16

Burn, Ernesto, 29

Butler, Frank, 174–5, 188–90

Café Martin, New York, 206

Cahill, Thaddeus, 206–7

Californian, SS, 249
Campania, SS, 194–5
Campbell-Swinton, A.A., 23
Canada: offers government support to GM, 103–6, 112
Cape Breton Island, Newfoundland: Canadian government offers site on to GM, 103–6; industrialisation, 144; wireless station on, 144; GM in, 159; larger transmitter built ('Marconi Tower'), 177, 185, 196; Beatrice visits, 195, 237; successful transmissions from, 205
Cape Cod, Massachusetts: wireless station established, 77–9, 86, 103, 112, 149, 157, 205; wireless mast collapses, 88–9
Cape Race, Newfoundland, 248–9
Carlo Alberto (Italian cruiser), 132, 134–41, 143, 145, 149
Carlton Hotel, London, 194
Carmania, SS, 265–6
Carpathia, SS, 244–5, 248–57
carrier pigeons: in World War I, 269–70
Carroll, Lewis (Charles Lutwidge Dodgson), 93
Caruso, Enrico, 228
Century Illustrated, The (monthly magazine), 104

Century Magazine, 28
Chelmsford, Essex, 76, 167–8, 283
Chicago Telephone Company, 215
Chiseltine, Captain, 85
Clarke, Russell, 271–2
Clerk Maxwell, James, 8, 14, 83, 97, 129
Cleveland, Grover, 108
Clifden, Ireland, 196, 205, 217, 227, 236
Clifton, Leopold Agar-Robartes, 5th Viscount, 72
Cobb Island, Potomac River, Maryland, 82–5
coherers, 17–19, 98, 122, 197, 239
Collier's Weekly, 109
Coltano, near Pisa, Italy, 160, 180, 260
Columbia (yacht), 61
Compagnie de Télégraphie Sans Fils, La (Brussels), 239
Consolidated Railway Telegraph Company, 63
Cook, Ed, 77–8
Cornwall, 71, 77
Cottam, Harold, 244–5, 248–9, 251, 253–7
Cragside (house), Northumberland, 22
Crescent (ship), 39
Crippen, Hawley Harvey, 231–5, 237, 243

Crookes, William, 96–7

Crookhaven, Ireland, 79

Crosby, Oscar T., 207

crystal set receivers, 204–5, 273

Cuba, 59, 188–90

Cunard Line (shipping), 64–5

Cunard, Thomas, 64

Daily Chronicle, 3

Daily Express (Dublin), 39–40

Daily Mail, 42, 191–3, 233–5, 283, 286

Daily Mirror, 191

Daily News (St John's), 112–14

Daily Telegraph, 266, 271, 290

Dam, H.J.W., 7–8

Darwin, Charles, 36

Davy, Sir Humphry, 13

daylight: effect on transmissions, 126, 135–6, 138, 209, 284, 286

Defence of the Realm Act (1914), 268, 270

de Forest, Lee: developments in wireless, 86, 152–7; activities in Russo–Japanese war, 160, 161–3, 165–6, 180; tower and transmissions at St Louis World's Fair, 172–5; kept from Fessenden's work, 187; sets up stations in Florida and Cuba, 188–9; abandoned by White, 190; devises speech and music receiver ('audion'), 207–8, 218, 259, 279; abandons rivalry with Marconi, 240; patents bought up, 283; obituary tribute to GM, 289–90

De Forest Wireless Telegraphy Company, 153–4

Dempsey, Jack, 115

Denison, Thomas S., 212

Deutsche Betriebsgesellschaft für Drahtlöse Telegraphie (DEBEG), 239

Deutschland (German liner), 136–7, 158

Dew, Detective Inspector Walter, 232–4

Dewey, Admiral George, 59–60

directional aerials, 196

Dolbear, Amos Emerson, 62, 109, 117

Dominion Steel and Coal Company, 105, 144

Doyle, Sir Arthur Conan, 44, 286; 'The Adventure of the Devil's Foot', 72

Dromoland Castle, Co. Clare, 182–3, 194

Ducretet, Eugène, 75, 176, 199n, 278

Dumas, Alexandre, 277

Duncan, HMS, 160

dynamite, 226

Eaglehurst (house),
  Hampshire, 240–1, 246, 274
earth: curvature, 53, 55
Edison, Thomas: Preece
  meets, 6; invents lightbulb,
  22; uses induction method,
  63; praises GM, 66, 291;
  Fessenden works with, 83–4;
  initial scepticism over
  transatlantic signals, 102–3;
  electrical fantasies, 110;
  absent from New York
  welcome for GM, 111; partial
  deafness, 143n; vacuum
  lightbulb effect, 207; and
  Puskas, 212
Edward VII, King (earlier
  Prince of Wales): uses early
  wireless link on royal yacht,
  41–2, 44; succeeds to throne,
  101; coronation postponed
  through illness and
  operation, 132–4, 137, 139,
  199; GM sends transatlantic
  greetings to, 146, 148
Edward VIII, King (later
  Duke of Windsor), 288
Egyptian Hall, Piccadilly,
  168
Eiffel, Gustave, 277
Eiffel Tower, Paris: as wireless
  mast, 277–9
Electric Club (USA), 63
Electric Girl Lighting
  Corporation, 109

Electrical Review, 215
Electrical World (US
  magazine), 66, 137
Electrician, The (magazine),
  129–30, 167, 170
electricity: for lighting, 21–2;
  early utilisation of, 108–10
Electro Importing Company,
  New York, 202
electro-magnetism: travels in
  waves, 8, 19, 96; early
  experiments in, 10, 14
Elena, Queen of Victor
  Emmanuel III, 260
Elettra (steam yacht), 284
Elmore, Belle (Mrs H.H.
  Crippen), 231–2
Escoffier, Georges-Auguste,
  194n
ether: as supposed transmission
  medium, 4, 56–7, 92, 126–7,
  287
Evening Herald (St John's,
  Newfoundland), 104
Evening Mail (Dublin), 40
Ewing, Sir Alfred, 271

Fahie, J.J.: History of Wireless
  Telegraphy 1838–1899, 97
Faraday, Michael, 13, 167
Fawley, Hampshire, 240
Fayant, Frank, 151
Ferrie, Captain Gustave,
  278
Fessenden, Helen, 82, 186

Fessenden, Reginald: as rival to GM, 75, 156–7; at Cobb Island, 82, 84; background and career, 82–4, 152; aims to transmit speech, 85, 208–9; Marriott uses system, 117; sues de Forest, 161; ambitions to send transatlantic signals, 186–7, 208–9; builds transmitter at Brant Rock, 186; writes to Heaviside, 187; de Forest applies to join, 190; sound broadcast on Christmas Eve 1906, 209–10; patents bought up, 283

Field, Kate, 25

Fitzgerald, George, 14

Fitzsimmons, Bob, 115–17

Flathold Island (Bristol Channel), 25–7

Fleming, Sir Ambrose: designs and operates Poldhu generator, 77–80, 130, 139, 142–3, 145, 149; described as GM's secretary, 79; believes in ether as waves medium, 126; lectures at Royal Institution, 167–8; devises valves based on vacuum lightbulbs, 207–8, 218, 273, 279; recommends GM for Nobel Prize, 226

Flood Page, Major Samuel see Page, Major Samuel Flood

Florida, SS, 220–3

Flying Huntress (tug), 39–41

Fortnightly Review, 96

France: first cross-Channel wireless messages transmitted, 45–8; see also Paris

Franklin, Benjamin, 13–14, 19, 197

Fraser, David, 162–4, 166; A Modern Campaign, 163

Fraser, Tim, 164

Frinton, Essex, 218n

Galvani, Luigi, 10

Gander, L. Marsland, 290

George V, King (earlier Prince of Wales), 158, 183, 236, 267, 270

Germany: develops Marconi's method, 29; hostility to Marconi Company, 136–9, 176, 185, 228, 239; submarine cables cut in World War I, 268; code system in World War I, 271–2; messages monitored and intercepted in World War I, 272–3

Gernsback, Hugo, 202–3

Given, Thomas H., 156, 210

Glace Bay, Nova Scotia: wireless station established, 141, 143–9, 157, 177, 196, 217; wireless structure collapses, 150

Gladstone, William Ewart, 37

Godalming: street lighting introduced, 22

Grande Duchesse, SS, 60

Great Eastern (ship), 89

Griffone, Villa, Pontecchio, near Bologna, 10, 12–16, 19–20, 236, 281–2, 288

Grosser Kurfürst, SS, 266

Guantanamo, Cuba, 188–9

Gurney, Edmund, 93

Haimun, SS, 164–6

Halifax Herald (Nova Scotia), 89–91, 105–6

Hall, Cuthbert, 159

Harmsworth, Alfred (*later* Viscount Northcliffe), 165

Harper's Weekly, 157, 223

Harrison, Benjamin, 108

Havel River, near Potsdam, 29

Haven Hotel, Poole, 37, 54, 68, 73, 159, 227

Heaviside, Charles, 129

Heaviside, Oliver: background and character, 128–30; theories on wireless telegraphy, 130–1, 228, 285; partial deafness, 143n; Fessenden makes offer to, 187; recommended for Nobel Prize, 228; death, 285

Heinrich, Prince of Germany, 137

Hertz, Heinrich, 7–8, 14–15, 96, 130, 258

Hertzian waves, xvi, 8, 15, 17–19, 24, 29, 31, 62–3, 96–8

Hibberd, A.S., 215

Hill, J. Arthur, 96

Hippisley, Richard, 271, 272

Hitchcock, Alfred, 234

Hohenzollern (German royal yacht), 137–8

Holman, Josephine: engagement to GM on *St Paul*, 67, 69–70, 184; absence from GM, 70, 80; breaks off engagement, 112–14

Howard de Walden, Thomas Evelyn Scott-Ellis, 8th Baron, 181, 183

Hozier, Colonel Sir Henry Montague, 176

Hughes, David Edward, 97–9

Hugo, Victor, 226

Hunstanton, Norfolk, 272

icebergs: as ocean hazard, 89, 242–3

Illustrated Mail, 73–4

Imperial Wireless Scheme, 263

Inch, Captain Francis, 264

Inchiquin, Edward Donough O'Brien, 14th Baron, 182–3, 194

Inchiquin, Ellen Harriet, Lady (*née* White), 184, 193, 195

Inchiquin, Lucius William
    O'Brien, 15th Baron
    (Beatrice's brother), 193
Irwin, Jack, 219–20
Isaacs, Godfrey, 239–40, 262,
    269
Isaacs, Harry, 262
Isaacs, Rufus (*later* 1st
    Marquess of Reading),
    262–3
Isle of Wight, xv, xvii, 35–6,
    39–43, 54–5, 66–8, 79
Italy: GM honoured in, 157,
    180, 261; war with Turkey
    (1911), 237–8; GM visits in
    wartime, 274; GM's political
    career in, 274, 281–2, 284,
    288; enters war (1915), 276;
    see also Griffone, Villa;
    Rome; Victor Emmanuel
    III, King

Jack the Ripper, 1
Jackson, Captain Henry, 31,
    160
James, Captain Lionel, 161–6,
    197
Jameson, Andrew (Annie's
    father), 10
Jameson-Davis, Henry (GM's
    cousin), 22, 32, 67
Jamison, Captain (of *St Paul*),
    69
Japan: war with Russia
    (1904–5), 161–6, 197, 198–201

Jeffries, Jim, 115–17
Journal of Commerce, 50

Kaiser Wilhelm der Grosse
    (German liner), 76, 136
Kemp, George: assists GM,
    40–1, 43, 47, 77–81, 85–6, 135,
    149; with GM in
    Newfoundland, 86–7, 89–91,
    100–1, 104; in New York, 108,
    110; on SS *Philadelphia*,
    121–2; prepares for Edward
    VII's coronation, 133; fits up
    equipment on *Carlo Alberto*,
    134; summoned to Kiel,
    137–8; van Raaltes visit, 179;
    on GM's motor car, 180, 185;
    keeps Beatrice and Degna
    amused in Poole, 227; helps
    refurbish yacht *Elettra*, 284
Kendall, Captain Harold, 231–4
Kennelly, Arthur, 131
Kiel, 137–8
Kingston Regatta, Dublin Bay,
    39–41, 44
Kipling, Rudyard, 44;
    'Wireless' (short story), 123
Kitchener, General Horatio
    Herbert (*later* 1st Earl), 162,
    269
kites and balloons: as aerial
    supports, 46, 86, 90–1,
    100–1
Knollys, Francis Knollys, 1st
    Viscount, 149

Kronprinz Wilhelm (German ship), 136–7
Kronstadt, 136
Kroonland, SS, 266

Lake Champlain, SS, 90
Langtry, Lillie (Emilie Charlotte le Breton), 37
Larnard, Lyman C., 62
Laurentic, SS, 233
Lavernock Point, Glamorgan, 25–7
Le Neve, Ethel, 231–5, 237
Lenin, Vladimir Ilyich, 2
Levitor kites, 46, 100–1
Liberté (French newspaper), 235
lightbeams: as message-senders, 108–9
Lipton, Sir Thomas, 61, 154
Liverpool: Marconi training school in, 218n
Livingstone, David, 60
Lizard Peninsula, Cornwall, 70–1, 77–9
Lloyd George, David: in 'Marconi scandal', 263
Lodge, (Sir) Oliver: scientific interests, 17, 23, 31–3, 75, 96, 98; interest in spiritualism, 91–2, 94–7, 286; scepticism over transatlantic signals, 102; belief in ether, 126; Commander Barber cites, 176; patent rights, 239–40;

retained by Marconi Company, 240; *Raymond*, 287
Lodge, Raymond, 286–7
Los Angeles Herald, 116
Los Angeles Times, 116
Louisberg, Cape Breton, 177
Lucania, SS, 158–60
Lusitania, SS, 246, 274–5; sunk by German U-boat, 275–6

McClure, Henry, 67, 69, 113, 124–6
McClure, Robert, 44, 46–8
McClure, Samuel, 44
McClure's (US magazine), 7, 44, 47, 52, 104, 124
Macdonald, Sir Claude, 165
McGrath, P.T., 104
Machrihanish, Mull of Kintyre, 187, 208–9
McKenzie, F.A., 165–6
Madeira Hotel, Bournemouth, 36–7
'Maggies' (magnetic receivers), 184, 197, 273
Maine, USS, 59
Majestic, SS, 161–2
Maltby, P.B., 265
Marconi, Alfonso (GM's brother): birth, 12; assists GM with early experiments, 19–20; music-making, 37, 54; as best man at GM's wedding, 193

Marconi, Annie (*née* Jameson; GM's mother): supports GM, 5; in London, 9, 20, 22; background, 11; marriage and children, 12, 45; on Isle of Wight, 36; in Poole, Dorset, 37, 73; concern for GM's comforts, 73; corresponds with GM, 185; and GM's marriage relations, 236; writes to GM on *Pisa*, 238; death, 284

Marconi, Beatrice (*née* O'Brien; GM's first wife; *later* Marignoli): meets and first rejects GM, 182–5; marriage to GM, 191–4; marriage relations, 194–5, 217, 227–8, 236–7, 260; accompanies GM to Glace Bay, 195, 217; birth and death of daughter, 195–6; birth of daughter Degna, 217; later pregnancy and birth of son, 227, 229; stays in Poole, 227; in Stockholm for GM's Nobel Prize award, 229; lives at Eaglehurst (Hampshire), 240–1; booked on *Titanic*'s maiden voyage, 245; in Italy, 260; in World War I, 274, 276; and birth of Gioia Jolanda, 276; divorce from GM and remarriage, 284–5;

objects to GM's sale of Rome home, 284; pays respects at GM's lying-in-state, 290

Marconi, Cristina (*née* Bezzi-Scali; GM's second wife), 285, 287–8, 290

Marconi, Degna (*later* Paresce; GM's daughter): and GM's parents' marriage, 11–12; on GM's cricket-playing, 80; on Inez Milholland, 159; birth, 217; childhood in Poole, 227; and mother's social life in London, 237; on life at Eaglehurst, 240; booked on *Titanic*'s maiden voyage, 245; and father's glass eye, 261; in World War I, 274, 276

Marconi, Domenico (GM's grandfather), 11

Marconi, Gioia Jolanda (GM's daughter): birth, 276

Marconi, Giulia (*née* de Renolis; GM's grandmother), 11

Marconi, Giulio (GM's son): at Eaglehurst, 241; booked on *Titanic*'s maiden voyage, 245; in World War I, 274, 276

Marconi, Giuseppe (GM's father): first marriage, 11; meets and marries Annie, 12, 45; and GM's early experiments, 16, 19; and funding of GM's patent rights, 30, 32; GM visits in Italy, 158, 180; death, 180, 193

Marconi, Guglielmo:
demonstrates at Toynbee Hall (1896), 3–5, 24; fluency in English, 4, 15, 43, 58, 73; encouraged by mother, 5, 20, 94; character and manner, 7, 40, 43, 54, 106, 193; early experiments, 7–9, 14–16, 18–19; birth and upbringing, 12–13; boyhood reading and studies, 13–15; fails naval entrance examination, 20; moves to London, 20–2; demonstrates at Lavernock–Flatholm Island, 27–8; financial backing for patent rights, 29–30, 32; ship-to-shore experiments on south coast, 36–8, 40–2; music-making, 37, 54; companies formed using name, 43n; uses Morse code, 53; in USA, 57, 58–65, 106, 108, 110–11, 158–9, 254; appearance, 59, 73, 193; as self-publicist, 66; engagement to Josephine Holman, 67; foreign rivals, 74–5; manufacturing companies, 76; builds station

at Cape Cod, 77–8; cricketing, 80; acquires motorcycle, 81; in Newfoundland, 86–7, 89–91, 93, 99–104; first hears transatlantic signals, 100–1; honoured in New York, 110–11; Josephine Holman breaks off engagement, 112–14; signals on SS *Philadelphia*, 121–6; sails on *Carlo Alberto*, 134–41, 143; manufactures magnetic coil detector, 135; in Kiel, 138; falls ill, 140, 196; home life, 140–1; and transatlantic signals between Poldhu and Glace Bay, 147–9; honoured in Italy, 157, 180, 261; engagement to Inez Milholland, 159, 177; security of signals breached, 169–70; visits St Louis World's Fair, 174; motoring, 180–1, 185; meets and falls in love with Beatrice O'Brien, 182–4; marriage to Beatrice, 191–4; jealousy over Beatrice, 194; later philandering, 194; birth and death of daughter, 195; finances, 195; birth of daughter Degna, 217; marriage relations, 217, 228, 236–7, 260; and SS *Republic* rescue, 223–4; shares Nobel

Marconi, Guglielmo – *cont.*
Prize for physics (1909), 228;
birth of son, 229; serves on
Pisa in Italo–Turkish war
(1911), 237–8; financial
success, 240; cancels trip on
Titanic, 246; and *Titanic*
disaster, 254–6; injured in
road accident, 260–1; glass
eye, 261; predictions about
political effects of wireless
technology, 261–2; and
shares scandal (1912–13),
262–3; granted honorary
GCVO, 267; political career
in Italy, 274, 281–2, 284, 288;
status in World War I, 274;
military service in World
War I, 275; birth of third
child (Gioia Jolanda), 276;
death, funeral and tributes,
281–2, 288–90; mausoleum,
281; post-war activities and
travels, 283–4, 287–8; buys
steam yacht, 284; separation
and divorce from Beatrice,
284–5; second marriage (to
Cristina), 285, 287;
Appleton's obituary tribute
to, 286; converts to
Catholicism, 288;
achievements, 289–91;
fortune and will, 290
Marconi, Luigi (GM's half-
brother), 12

Marconi, Maria Elettra Elena
Anna (GM/Cristina's
daughter), 287–8, 290
Marconi House, London, 218n
Marconi International Marine
Company: set up, 43n, 76;
expansion and operations,
184
Marconi Wireless Telegraph
and Signal Company: set
up, 32–3; GM reports to on
US reception, 58; German
hostility to, 136–9, 176, 185,
228, 239; contract with Royal
Navy, 160; and de Forest in
Russo–Japanese war, 165;
withdraws from St Louis
World's Fair, 173; expansion,
184; training schools, 218,
243; prospers, 238–9;
association with Deutsche
Betriebsgesellschaft, 239; and
shares scandal (1912–13), 263;
taken over by British
government in Great War,
270–1, 283; manufactures
radio receivers, 283
Marconiograph (company
magazine), 260–1
Maria Theresa, Archduchess
of Austria, 284
Marignoli, Marchesa Beatrice
see Marconi, Beatrice
Marignoli, Marchese Liborio:
Beatrice marries, 284

Marriott, Robert H., 116–17

Marsden, Chief Officer (of *Philadelphia*), 123, 125

Mary, Queen of George V (*earlier* Princess of Wales), 158, 236

Maskelyne, Nevil, 168–71

Masterman, Charles Frederick Gurney: *The Condition of England*, 182

Maupassant, Guy de, 277

Maxwell, James Clerk *see* Clerk Maxwell, James

May Flower (coastal steamer), 36

Melba, Dame Nellie, 283

Merrick, Joseph ('the Elephant Man'), 133

Milholland, Inez (*later* Boissevain): engagement to GM, 159, 177; GM breaks off engagement, 184; travels to Europe in First World War, 276

Mills, Captain A.R., 122–5

Mirabello, Admiral Carlo, 133, 137–40

Moffett, Cleveland, 44–7, 54

Monet, Claude, 21

Montrose, SS, 230–3

Morgan, John Pierpont, 243

Morning Leader, 170

Morse, Samuel Finley Breese, 49–51

Morse, Sidney, 50

Morse code, 16–17, 49–53

Motor Car Act (1904), 181

motor cars: beginnings, 180–2

Muirhead, Alexander, 96, 239–40

Mull of Kintyre, 187

Mullion, Cornwall, 71–3, 85; *see also* Poldhu Hotel

Mullis, P.R., 23

Murray (of Elibank), Alexander William Charles Oliphant Murray, 1st Baron, 263

Murray, Dr Erskine, 37, 54

Murray, George, 105

Mussolini, Benito: commissions mausoleum for GM, 281; rise to power, 284; invades Abyssinia, 288; orders state funeral for GM, 288, 290

Myers, Frederic, 93–4, 287

Nantucket, 121

Narragansett (oil tanker), 265–6

National Electric Signalling Company (NESCO), 156, 190, 209, 240

New England Electronic Music Company, 206

New York: GM visits, 57, 58–65, 106, 108, 110–12, 159, 254

New York Electronic Music Company, 207

New York Herald: GM covers America's Cup races for, 57, 83, 102; reports Admiral Dewey's homecoming, 60; publicises GM, 61, 102–3; uses searchlight for messages, 108; and German hostility to GM, 137; and *Titanic* story, 256

New York Times: reports GM's successes, 61; on New York reception for GM, 110–11; and GM's differences with Germans, 137; on de Forest's broadcasting claims, 208; reports *Republic* disaster, 223; and *Titanic* story, 254–7; on Sayville station passing on information in Great War, 275; and US technological development in Great War, 279; obituary tribute to GM, 289

New York Times Magazine, 204

New York Tribune, 59, 276

New York World, 108

Newfoundland, 86–91, 93, 99–104, 147

Niagara Falls, 22, 82, 109

Nicholas II, Tsar of Russia, 134, 136, 198–9

Niton, Isle of Wight, 79

nitro-glycerine, 225–6

Nobel, Alfred, 225–6

Nobel, Immanuel, 225

Nobel Foundation: established, 226; awards physics prize to GM, 228

Northcliffe, Alfred Harmsworth, Viscount *see* Harmsworth, Alfred

O'Brien, Beatrice *see* Marconi, Beatrice

O'Brien, Lilah (Beatrice's sister), 183, 229, 274

Olympic, SS, 244, 246, 250–3

Osborne (royal yacht), 39, 53, 149

Osborne House, Isle of Wight, 25, 35, 39, 42, 132

Osman, Lieut.-Colonel A.H.: *Pigeons in the Great War*, 270

Ottawa, 105–6

Page, Major Samuel Flood, 67–8, 72

Paget, Percy, 87, 88, 90–1, 100–1, 104

Palladino, Eusapia, 96–7

Paris: in World War I, 278; *see also* Eiffel Tower

Parkin, George, 147–8

Pennington, John, 264–5

Philadelphia, SS, 114, 121–6, 147

Philippines: US naval action in, 59

Phillips, Jack, 244–5, 247–51, 254–5, 257

Philosophical Magazine, 129
pigeons (homing) *see* carrier
    pigeons
Piper, Leonore, 94–5
Pisa, Italy *see* Coltano
Pisa (Italian warship), 237–8
Poldhu Hotel, Mullion,
    Cornwall: wireless station
    established, 72–4, 77–81, 86,
    89–91, 177; signals heard
    across Atlantic, 100–1;
    station communicates with
    SS *Philadelphia*, 122–5;
    communicates with *Carlo
    Alberto*, 133, 136, 139–40;
    King Victor Emmanuel
    visits, 139; transmitter power
    and discharges, 142–3, 145;
    communicates with Glace
    Bay, 146–9, 289; Prince and
    Princess of Wales visit, 158;
    messages intercepted, 170
Ponce, SS, 60–1
Poole, Dorset, 37, 43, 54, 73–4,
    178
Popov (Popoff), Alexander, 75,
    199n
Popov-Ducretet wireless
    system, 199
Port Arthur, 162, 164
Post Dispatch (St Louis), 118,
    172
Post Office (British): Preece
    works for, 3–4, 6, 23–4, 26;
    loses patent rights to

Marconi company, 33;
    monopoly on telegraphic
    messaging, 146; issues
    licences for wireless
    telegraphy, 185; licenses
    wireless, 283
Poulain, M. & Mme. (of
    Haven Hotel, Poole), 178
Preece, Sir William:
    encourages young GM, 2–5,
    7, 31, 43, 94, 97, 218, 291;
    induction process, 3, 62;
    background and career, 5–6;
    GM first meets, 23–4;
    experiments at Flatholm
    Island–Lavernock, 24–8;
    demonstrates Bell's
    telephone to Queen
    Victoria, 26; and Slaby,
    27–9, 136; and GM's
    forming commercial
    company, 33; Heaviside
    criticises, 129–30;
    Commander Barber cites,
    176
Prescott, General (GM's
    uncle), 12
Pulitzer, Joseph, 108
Punch (magazine), 267
Puskas, Ferenc, 212
Puskas, Tivador, 212–13

Raalte, Charles van, 178, 193
Raalte, Florence van, 178, 193
Raalte, Margarite van, 178–9

Raalte, Nony van, 179
radio: first broadcast (by
    Fessenden, 1906), 210;
    development of
    broadcasting, 280, 282
Radio Broadcast (magazine),
    174
Radio Corporation of America
    (RCA), 283
Raineri-Biscia (Italian
    midshipman), 135–6
Reading, 1st Marquess of *see*
    Isaacs, Rufus
Reith, John (*later* Baron), 287
Renolis, family de, 11
Republic, SS, 219–24
Richmond Park, 236
Righi, Augusto, 15, 16
Ritz, César, 194n
Roanoke Island, North
    Carolina, 85
Roentgen, Wilhelm Conrad:
    discovers X-rays, 6–7, 83, 95;
    awarded first Nobel Prize
    for physics, 227
Rome: GM made citizen of,
    157–8; GM lives in, 288
Roosevelt, Theodore, 149
Rostron, Captain Henry,
    249–51, 253
Round, H.J., 236, 273
Royal Cornwall Gazette, 79
Royal Institution, London,
    167–9
Royal Navy: contract with
    Marconi Company, 160; in
    Great War, 270
Royal Needles Hotel, Isle of
    Wight, xv, 36–7, 39, 54,
    66–7, 69
Rozhestvensky, Admiral
    Zinoviy Petrovich, 198–201
Rubbi, Luigi, 261
Russia: war with Japan
    (1904–5), 161–6, 197, 198–9;
    Tsushima defeat, 200–1
Rutherford, Ernest (*later*
    Baron), 134

St John's, Newfoundland,
    88–90, 101–4
St Louis, Missouri: World's
    Fair (1904), 172–5
St Paul, SS, 64–9, 275
Sammis, W.T., 256
Santa Catalina Island,
    California, 115–17
Sardinian, SS, 88–9, 104
Sarnoff, David, 290
Saturday Evening Post, 151
Savannah, SS, 50
Sayville wireless station, Long
    Island, 275
Schieger, Captain Walther, 275
Science Siftings (journal), 109
Scientific American (magazine),
    202, 205, 213–14
Scribner's Magazine, 123
Sealby, Captain Inman, 220–1,
    223

Searle, George F.C., 128–9
Sedden, Walter, 264–5
'Sellers, Colonel' (fictional character), 151–2
semaphore stations, 17
Shamrock (yacht), 61
Shamrock III (yacht), 154
Shantung, China, 163–4
Shaw, George Bernard, 95
Sheardown, Lucile: de Forest marries and separates from, 189–90
Shinamo Maru (Japanese cruiser), 200
Siasconset station, Nantucket, 219–21
Siberia, SS, 162–3
Signal Hill, Newfoundland, 90–1, 93, 100
Slaby, Adolphus, 27–9, 74, 136–7, 139, 185
Smith, Captain Edward, 244, 247–9
Smith, William, 103
Society for Psychical Research, 93–4, 287
Solari, Luigi, 132–3, 136, 138–40, 158–9, 165
Solent (coastal steamer), 36
SOS: as international distress signal, 222, 250
South Wellfleet Hotel, Cape Cod, 78
Spanish–American war (1898), 59–60

speech: wireless transmission of, 278–9
spiritualism, 92, 94–5, 286–7
Stanley, (Sir) Henry Morton, 60
steamships: on Atlantic route, 64
Steinbjel, Professor, 2
Steinmetz, Charles Proteus S., 111
Stevenson, Robert Louis, 37
Strand Magazine, 7
Stubblefield, Bernard, 118
Stubblefield, Nathan B., 118–19
Success (US magazine), 151, 155
Sunbourne (house), Hampshire, 217
Sutter, Countess Bertha (*née* Kinsky), 226
Swan, Joseph, 22
Swedish Academy, 226–7
Sydney, Nova Scotia, 105, 144, 149

Taft, Wiliam, 254
Technical World (magazine), 261
Telconia (cable-laying ship), 268
Telefon Hirmondo, Budapest, 211–14, 216
Telefunken (German company), 184, 239
telepathy, 94, 287
telephone broadcasting systems, 211–16

Telharmonium
(Dynamaphone), 206–7
'Telimco Wireless Telegraph
Outfit', 202
Tennyson, Alfred, 1st Baron,
93
Tesla, Nikola, 109, 111, 126,
212
Thumb, 'General' Tom, 168n
Times, The: on Edward VII's
coronation postponement,
133; reports GM's
transatlantic signals, 147–8;
misreports birth of Vyvyan's
daughter, 150; James reports
on Russo–Japanese war for,
162–3, 165; and Maskelyne's
Royal Institution prank, 169
Titanic, SS, 230n, 244–51,
253–4, 257–60
Togo, Admiral Count
Heihachiro, 200
Torquay, Devon, 128–9
Totland Bay Post Office, 69
Toynbee Hall, London, 1–2,
24, 33, 218
Transatlantic Times, 76
Treves, Sir Frederick, 133–4,
139
Trevithick, Richard, 73
Tsushima Straits, Battle of
(1905), 198–201
Tuckerton, New Jersey, 275
tuning (and wavebands), 75–6,
140

Turkey: war with Italy (1911),
237–8
Twain, Mark: creates Colonel
Sellers character, 151–2

U-20 (German submarine), 275
Umberto, King of Italy, 132
Umbria, SS, 125
United Fruit Company, 175, 209
United States of America: GM
first visits, 57–62; GM
revisits with Kemp, 106, 108,
210–12; first working
wireless station (California),
116; Prince Heinrich visits,
137; direct transmission of
messages across Atlantic,
149–50; regulates wireless
use, 157–8; GM revisits
(1903), 159; amateur and
recreational interest in
wireless, 202–5, 267–8, 280;
telephone broadcasting
systems, 215–16; technical
improvements in wireless
transmission, 259; neutrality
in World War I, 275–6;
technological supremacy in
World War I, 279–80, 282;
development of radio
broadcasting in, 282–3
United States Navy: considers
Marconi's wireless system,
61–2; encourages home
talent, 84; contracts with de

Forest, 175–6, 188, 190; indecision over adopting wireless system, 201; installs receivers on ships, 209; invests in new technology in World War I, 280; seeks to control all wireless in USA, 283

United States Weather Bureau, 83–5, 156

United Wireless (company), 217, 240

Ural (Russian warship), 200

Vail, Alfred, 50–1, 135

Vail, George, 51, 135

Vail, Stephen, 50, 135

Vail, Theodore N., 279

valves (radio): devised, 207–8, 218, 279; superiority of receivers, 273

Vanity Fair (magazine), 192

Verlaine, Paul, 277

Victor Emmanuel III, King of Italy: and Edward VII's postponed coronation, 132, 134; visits Russia and Germany, 136–7; GM signals to on *Carlo Alberto*, 139–40; visits Poldhu, 139; GM sends transatlantic greetings to, 146, 148–9; introduces GM to Kaiser, 158; honours GM, 180; message to GM on marriage

to Beatrice, 193; invites GM and Beatrice to home, 260

Victoria, Queen: receives text messages, xvii, 39, 42, 44; Alexander Graham Bell demonstrates to, 25; holidays, 35; GM granted audience with, 42; death, 132

Vladivostok, 198–201

Volta, Alessandro, 10

Volturno, SS, 263–6

Vyvyan, Jane, 150, 159, 195

Vyvyan, Richard: as GM's engineer, 72, 77–8, 89, 144–6; marriage and child, 150; and restrictions on Beatrice in Nova Scotia, 217

Walker, Hay, Jr, 156, 187, 190, 210

Washington Times, 119

wavebands *see* tuning

wavelength, 258, 285

Wei-hai-Wei, China, 164–6

Wheatstone, Charles, 128–9

White, Abraham: character and career, 151–2; backs and employs de Forest, 152–7, 162, 174; coup in Russo–Japanese War, 156, 162, 166; Maskelyne and, 171; publicity at St Louis World's Fair, 173, 175, 189; breaches Fessenden patents, 190; forms United Wireless company, 217

White Star Line (shipping), 243–5

Wilhelm II, Kaiser, 29, 136–8, 139, 158, 199

Willenborg, William J., 204

Wimereux, near Boulogne-sur-Mer, France, 45–7

Wireless Act (1905), 203

Wireless News (magazine), 154

Wireless Telegraph and Signal Company *see* Marconi Wireless Telegraph and Signal Company

wireless telegraphy: defence dangers, 9; Hughes discovers, 97–8; first use in apprehending criminals, 117–18; as means of mass communication, 120; affected by daylight, 126, 135–6, 138, 209, 284, 286

World War I, 268–71; monitoring stations established, 272–3; wireless technology in, 279–80, 282

'World-Wide Wireless', 154

World's Work (magazine), 212

X-rays, xvi, 6–7, 83, 95, 227

Yellow Sea, 162–3, 165, 174

9 780007 130061